T0206070

SpringerBriefs in Statistics

JSS Research Series in Statistics

Editors-in-Chief

Naoto Kunitomo
Akimichi Takemura

Series editors

Genshiro Kitagawa
Tomoyuki Higuchi
Toshimitsu Hamasaki
Shigeyuki Matsui
Manabu Iwasaki
Yasuhiro Omori
Masafumi Akahira
Takahiro Hoshino
Masanobu Taniguchi

The current research of statistics in Japan has expanded in several directions in line with recent trends in academic activities in the area of statistics and statistical sciences over the globe. The core of these research activities in statistics in Japan has been the Japan Statistical Society (JSS). This society, the oldest and largest academic organization for statistics in Japan, was founded in 1931 by a handful of pioneer statisticians and economists and now has a history of about 80 years. Many distinguished scholars have been members, including the influential statistician Hirotugu Akaike, who was a past president of JSS, and the notable mathematician Kiyosi Itô, who was an earlier member of the Institute of Statistical Mathematics (ISM), which has been a closely related organization since the establishment of ISM. The society has two academic journals: the Journal of the Japan Statistical Society (English Series) and the Journal of the Japan Statistical Society (Japanese Series). The membership of JSS consists of researchers, teachers, and professional statisticians in many different fields including mathematics, statistics, engineering, medical sciences, government statistics, economics, business, psychology, education, and many other natural, biological, and social sciences.

The JSS Series of Statistics aims to publish recent results of current research activities in the areas of statistics and statistical sciences in Japan that otherwise would not be available in English; they are complementary to the two JSS academic journals, both English and Japanese. Because the scope of a research paper in academic journals inevitably has become narrowly focused and condensed in recent years, this series is intended to fill the gap between academic research activities and the form of a single academic paper.

The series will be of great interest to a wide audience of researchers, teachers, professional statisticians, and graduate students in many countries who are interested in statistics and statistical sciences, in statistical theory, and in various areas of statistical applications.

More information about this series at http://www.springer.com/series/13497

Shuhei Mano

Partitions, Hypergeometric Systems, and Dirichlet Processes in Statistics

 Springer

Shuhei Mano
The Institute of Statistical Mathematics
Tachikawa, Tokyo, Japan

ISSN 2191-544X   ISSN 2191-5458   (electronic)
SpringerBriefs in Statistics
ISSN 2364-0057   ISSN 2364-0065   (electronic)
JSS Research Series in Statistics
ISBN 978-4-431-55886-6   ISBN 978-4-431-55888-0   (eBook)
https://doi.org/10.1007/978-4-431-55888-0

Library of Congress Control Number: 2018947483

Printed on acid-free paper

This Springer imprint is published by the registered company Springer Japan KK part of Springer
Nature
The registered company address is: Shiroyama Trust Tower, 4-3-1 Toranomon, Minato-ku, Tokyo
105-6005, Japan

Preface

This monograph is on statistical inferences related to some combinatorial stochastic processes. Especially, it discusses the intersection of three subjects that are studied almost independently from each other: partitions, hypergeometric systems, and Dirichlet processes. It is shown that the three subjects involve a common structure called exchangeability. Then, inference methods based on their algebraic nature are presented. The topics discussed are rather simple problems, but it is hoped that the present interdisciplinary approach attracts a wide audience.

I am indebted to many people in my studies of the subjects discussed in this monograph and in preparing this monograph. I would like to express my gratitude to Prof. Akimichi Takemura, the editor of this volume, for his continuous encouragement throughout the preparation of this work. Professor Akinobu Shimizu and Prof. Koji Tsukuda read a draft of this work and gave careful and useful comments. I would like to thank Mr. Yutaka Hirachi of Springer Japan for his help and patience. This work was supported in part by Grants-in-Aid for Scientific Research 15K05013 from Japan Society for the Promotion of Science.

Tokyo, Japan   Shuhei Mano
May 2018

Contents

Chapter 1
Introduction

Abstract Partitions appear in various statistical problems. Moreover, stochastic modeling of partitions naturally assumes the exchangeability. This chapter introduces this monograph by providing minimum definitions and terminologies related to partitions and exchangeability. To illustrate the content of this monograph, we present two simple statistical problems involving partitions.

Keywords A-hypergeometric distribution · Algebraic statistics
Bayesian nonparametrics · Dirichlet process · Exchangeability · Partition

1.1 Partition and Exchangeability

In this monograph, we discuss some combinatorial stochastic processes, which play important roles in Bayesian nonparametrics and appear in count data modeling. The statistical inferences inevitably involve two algebraic structures: partition and exchangeability. This section defines both structures and introduces their fundamental concepts.

A *partition* is defined as follows.

Definition 1.1 A partition of a finite set F into k subsets is a collection of non-empty and disjoint subsets $\{A_1, \ldots, A_k\}$ whose union is F.

Consider a sequence

$$\lambda = (\lambda_1, \lambda_2, \ldots)$$

of nonnegative integers in decreasing order $\lambda_1 \geq \lambda_2 \geq \cdots$ and containing finitely many nonzero terms. The nonzero λ_i are called the *parts* of λ. The number of parts is called the *length* of λ, denoted by $l(\lambda)$, and the sum of parts is called the *weight* of λ, denoted by

$$|\lambda| := \lambda_1 + \lambda_2 + \cdots + \lambda_{l(\lambda)}.$$

S. Mano, *Partitions, Hypergeometric Systems, and Dirichlet Processes in Statistics*,
JSS Research Series in Statistics, https://doi.org/10.1007/978-4-431-55888-0_1

If $|\lambda| = n$, λ is said to be a partition of a positive integer n, or an *integer partition* of n, denoted as $\lambda \vdash n$. For convenience, the notation

$$c_i(\lambda) := \#\{j; \lambda_j = i\}, \qquad i \in [n] := \{1, 2, \ldots, n\}$$

is sometimes used, indicating the number of times each integer occurs as a part in λ. In this expression, $\#\{A\}$ represents the cardinality of set A. In other words, c_i is the multiplicity of i in λ. Because multiplicities of integer partitions will frequently appear in this monograph, we specifically refer to the set of multiplicities as *size index*, a term coined by Sibuya [1].

We now define exchangeability. In this monograph, we explore only a few aspects of exchangeability. An extensive survey on exchangeability was given by Aldous [2]. We write $X \overset{d}{=} Y$ if random variables X and Y have the same distribution.

Definition 1.2 If a finite sequence of random variables $(X_1, \ldots, X_n)$ satisfies

$$(X_1, \ldots, X_n) \overset{d}{=} (X_{\sigma(1)}, \ldots, X_{\sigma(n)})$$

for any permutation σ of $[n]$, then $(X_1, \ldots, X_n)$ is called *n-exchangeable* (n is used to indicate the finite number of random variables). If an infinite sequence $(X_1, X_2, \ldots)$ satisfies

$$(X_1, X_2, \ldots) \overset{d}{=} (X_{\sigma(1)}, X_{\sigma(2)}, \ldots)$$

for any finite permutation σ ($\#\{i; \sigma(i) \neq i\} < \infty$) of $\mathbb{N} := \{1, 2, \ldots\}$, then $(X_1, X_2, \ldots)$ is called *exchangeable*.

Example 1.1 Consider that there are n balls in an urn, each ball labeled with a distinct number in $[n]$. When the balls are chosen and replaced, the infinite sequence of drawn numbers is exchangeable. When the balls are drawn and not replaced, the finite sequence of n numbers is n-exchangeable.

An independent and identically distributed (i.i.d.) sequence of random variables is exchangeable, but the converse is not always true. In fact, exchangeability restricts the correlation structure. The correlation coefficients of two random variables in an n-exchangeable sequence $(X_1, \ldots, X_n)$ satisfy $\rho(X_i, X_j) \geq -1/(n-1)$, $\forall i \neq j$. In Example 1.1 the lower bound saturates when the balls are drawn without replacement.

A probability measure on partitions is called a *random partition*. Let us introduce *infinite exchangeability* of random partitions. For further discussion on this subject, see Chap. 11 of [2] and Chap. 2 of [3]. Let us call an unordered integer partition *composition*, and $\mathscr{C}_n$ denote the set of all compositions of n. A random partition Π_n of a finite set $[n]$ is called exchangeable if its law is invariant under any permutation of $[n]$. Equivalently, for each partition $\{A_1, \ldots, A_k\}$ of $[n]$,

$$\mathbb{P}(\Pi_n = \{A_1, \ldots, A_k\}) = p_n(|A_1|, \ldots, |A_k|)$$

for some symmetric function p_n of compositions $(n_1, \ldots, n_k) \in \mathscr{C}_n$. This function p_n is called the *exchangeable partition probability function* (EPPF) of Π_n. For $m \in [n]$, let $\Pi_{m,n}$ denote the restriction of Π_n to $[m]$. Then, $\Pi_{m,n}$ is an exchangeable random partition of $[m]$ with some EPPF $p_n : \mathscr{C}_m \to [0, 1]$. So for each partition $\{A_1, \ldots, A_k\}$ of $[m]$

$$\mathbb{P}(\Pi_{m,n} = \{A_1, \ldots, A_k\}) = p_n(|A_1|, \ldots, |A_k|)$$

is extended recursively to $\mathscr{C}_m$ for $m = n - 1, n - 2, \ldots, 1$, using the addition rule of probability. For each composition $(n_1, \ldots, n_k)$ of $m < n$

$$p(n_1, \ldots, n_k) = p(n_1, \ldots, n_k, 1) + \sum_{i=1}^{k} p(n_1, \ldots, n_i + 1, \ldots, n_k). \qquad (1.1)$$

A sequence of exchangeable random partitions $(\Pi_1, \Pi_2, \ldots)$ is called *consistent*[1] in distribution if Π_m has the same distribution as $\Pi_{m,n}$ for every $m < n$. Equivalently, there is a symmetric function p defined on the set of all integer compositions such that $p_1(1) = 1$, the addition rule (1.1) holds for all integer compositions, and the restriction of p to $\mathscr{C}_n$ is the EPPF of Π_n. Such a sequence $(\Pi_1, \Pi_2, \ldots)$ can be constructed so that $\Pi_m = \Pi_{m,n}$ almost surely for every $m < n$. The sequence of random partitions $\Pi_\infty := (\Pi_1, \Pi_2, \ldots)$ is then called an *infinite exchangeable random partition*.

1.2 Examples

This section presents two examples wherein exchangeable partitions appear. They are intended as representative problems to be discussed throughout this monograph. The former problem appears in Bayesian nonparametrics, and the latter appears in algebraic statistics.

1.2.1 Bayesian Mixture Modeling

The following Bayesian mixture model is a simplified version of the Dirichlet process mixture model discussed by Lo [5]. Section 6.3 of [6] summarizes the recent computational developments concerning the Dirichlet process mixture models. Consider a clustering of a sample of size n. Let us fix the number of clusters $k = 2$, and let the cluster that the i-th observation belongs to be $x_i \in \{1, 2\}$. A clustering model

---

[1] This property is referred by several names in the literature. In the original paper by Kingman [4], it was called the *partition structure*.

assumes that each observation independently belongs to the first and second clusters with probabilities θ and $1 - \theta$, respectively. Then, the sequence $(X_1, \ldots, X_n)$ is a Bernoulli trial of parameter θ. Suppose we have a sample $(Y_1, \ldots, Y_n)$ and assume a normal mixture model

$$Y_i | X_i \sim \mathrm{N}(\mu_{X_i}, \sigma^2), \qquad X_i \sim \mathrm{Ber}(\theta), \qquad i \in [n].$$

To simplify the discussion, we assume that the parameters of the normal distribution are known. Then, a clustering is a sample from the posterior distribution

$$\mathbb{P}(X | Y; \theta) \propto \prod_{i=1}^{n} \phi(Y_i | X_i, \sigma) \mathbb{P}(X; \theta),$$

where $\phi(Y_i | X_i, \sigma)$ is the normal density of mean μ_{X_i} and variance σ^2, and

$$\mathbb{P}(X_1 = x_1, \ldots, X_n = x_n; \theta) = \theta^t (1 - \theta)^{n-t}. \tag{1.2}$$

Here, the size of the first cluster $T(X) := \#\{i; X_i = 1\}$ is the sufficient statistic of the Bernoulli trial.

In the Bayesian context, we may take a mixture over the parameter θ. Let us take the beta distribution (the conjugate prior to the Binomial distribution) as the mixture distribution. For the beta distribution of parameters α and β, we have

$$\mathbb{P}_{\alpha,\beta}(X_1 = x_1, \ldots, X_n = x_n)$$
$$= \int_0^1 \mathbb{P}(X_1 = x_1, \ldots, X_n = x_n; \theta) F_{\alpha,\beta}(d\theta) = \frac{(\alpha)_t (\beta)_{n-t}}{(\alpha + \beta)_n}, \tag{1.3}$$

where

$$F_{\alpha,\beta}(d\theta) = \frac{\Gamma(\alpha + \beta)}{\Gamma(\alpha)\Gamma(\beta)} \theta^{\alpha-1}(1 - \theta)^{\beta-1} d\theta \tag{1.4}$$

and we have used the rising factorial $(x)_i := x(x + 1) \cdots (x + i - 1)$. This probability mass function is an example of an EPPF. The distribution of $T(X)$ given by

$$\mathbb{P}_{\alpha,\beta}(T(X) = t) = \binom{n}{t} \frac{(\alpha)_t (\beta)_{n-t}}{(\alpha + \beta)_n} \tag{1.5}$$

is called the beta-binomial or negative hypergeometric distribution.

Remark 1.1 If the sequence of labels $(X_1, \ldots, X_n)$ were observable, the expression (1.3) would be the *marginal likelihood* in Bayesian terminology, as we integrate over the parameter θ of the likelihood (1.2) with the prior distribution (1.4). In reality, the labels are unobservable, and the marginal likelihood of the model is

$$\mathbb{P}_{\alpha,\beta}(Y) = \sum_x \prod_{i=1}^n \phi(Y_i | X_i = x_i, \sigma) \mathbb{P}_{\alpha,\beta}(X = x).$$

In an empirical Bayes procedure, the estimated hyperparameters α and β should maximize the marginal likelihood.

Note that (1.3) implies that the marginal distribution of the sequence $(X_1, \ldots, X_n)$ is exchangeable, but not i.i.d. The expression (1.3) represents the exchangeable sequence as a mixture of i.i.d. sequences. This modeling is justified by the celebrated de Finetti's representation theorem.

Theorem 1.1 (de Finetti 1937 [7]) *An infinite exchangeable sequence of random variables is a mixture of i.i.d. sequences of random variables.*

Informally, de Finetti's representation theorem states that an exchangeable sequence is an i.i.d. sequence of a random probability measure, sometimes called the *de Finetti measure*. If the observed sequence is sufficiently long, we can infer the realized (true) probability measure. In the above modeling, the random probability measure Λ on labels of clusters is given by

$$\Lambda := \theta \delta_1 + (1 - \theta) \delta_2. \tag{1.6}$$

As θ is a beta random variable, Λ should be a random probability measure. Suppose that the realized sequence is a Bernoulli trial with parameter θ_0. By the strong law of large numbers, $T/n \to \theta_0$, *a.s.* and the empirical measure is

$$\Lambda_n(X_1, \ldots, X_n) := \frac{T}{n} \delta_1 + \left(1 - \frac{T}{n}\right) \delta_2 \to \Lambda_0, \qquad a.s.$$

as $n \to \infty$, where Λ_0 is the realized probability measure given by replacing θ with θ_0 in (1.6). By conjugacy, we have the posterior measure

$$\mathbb{P}_{\alpha,\beta}(X_{n+1} = \cdot | X_1, \ldots, X_n) = \frac{\alpha + \beta}{\alpha + \beta + n} \Lambda_* + \frac{n}{\alpha + \beta + n} \Lambda_n,$$

$$\Lambda_* := \frac{\alpha}{\alpha + \beta} \delta_1 + \frac{\beta}{\alpha + \beta} \delta_2.$$

This expression is a convex combination of the prior measure Λ_* and the empirical measure Λ_n. As $n \to \infty$, the posterior measure converges to the realized measure Λ_0 almost surely.

The above discussion is extendable to any number of clusters. For multiple clusters, the role played by the beta distribution is played by the Dirichlet distribution. The support of the random probability measure with an m-variate Dirichlet distribution is $[m]$. Ferguson [8] defined an analogous object with a support of $\mathbb{R}$, which is now called Ferguson's Dirichlet process.

Definition 1.3 (*Ferguson 1973* [8]) Let α be a finite measure on a measurable space $(\mathbb{R}, \mathscr{B}(\mathbb{R}))$, where $\mathscr{B}(\mathbb{R})$ is a Borel field on $\mathbb{R}$. If a random probability measure F on $\mathbb{R}$ satisfies

$$(F(A_1), \ldots, F(A_k)) \sim \mathrm{Dir}(\alpha(A_1), \ldots, \alpha(A_k))$$

for any finite partition $\{A_1, \ldots, A_k\}$ of $\mathbb{R}$, then F is called a Dirichlet process.

The Dirichlet process is a fundamental prior process in Bayesian nonparametrics. Chapter 4 will introduce the Dirichlet and related processes relevant to our discussion. The Dirichlet process is probably most commonly used in mixture modeling. However, this monograph mainly discusses EPPFs, which are marginal likelihoods if their labels are observed (Remark 1.1), as in some categorical data. Focusing on EPPFs will simplify our discussion. Chapter 5 will discuss samplers and conditional maximum likelihood estimation involving EPPFs.

1.2.2 Testing Goodness of Fit

Diaconis et al. [9] discussed a goodness-of-fit test for a Poisson regression. Consider a discrete covariate with m levels, and let the means of independent Poisson random variables C_i be $\mu_i = \exp(\alpha + \beta i)$, $i \in [m]$. Also let $\hat{\alpha}$ and $\hat{\beta}$ be the maximum likelihood estimates of α and β, respectively. A goodness-of-fit test of the Poisson regression model with $\hat{\mu}_i = \exp(\hat{\alpha} + \hat{\beta} i)$ can be constructed based on the chi-square statistic

$$\chi^2 = \sum_{i=1}^{m} \frac{(\hat{\mu}_i - C_i)^2}{\hat{\mu}_i}.$$

Asymptotic theory predicts an approximate chi-square distribution with $m - 2$ degrees of freedom. However, suppose we have a small sample that may not be analyzed by asymptotic theory. The statistics

$$K(C) := C_1 + \cdots + C_m, \qquad N(C) := 1 \cdot C_1 + 2 \cdot C_2 + \cdots + m \cdot C_m \quad (1.7)$$

are sufficient for the parameters α and β, where $K(C)$ and $N(C)$ are the total numbers of counts and levels, respectively. Let us assume a sample c_* with $K(c_*) = k$, $N(c_*) = n$, and a chi-squared value of χ_*^2. Consider a sample of size k comprising independent label observations. Let the level of the i-th observation be $N_i \in [m]$. Then, the count vector $(N_1, \ldots, N_k)$ is k-exchangeable with the sum of n, which determines a partition of n with length k, and $C_i = \#\{j; N_j = i\}$ is the multiplicity of i. Define the support $\mathscr{S}_{n,k}$ of a count vector $(C_1, \ldots, C_m)$ by (1.7), the set of all nonnegative m-tuples $(c_1, \ldots, c_m)$ with $K(c) = k$ and $N(c) = n$. Obviously, the problem is combinatorial, so that the cardinality of $\mathscr{S}_{n,k}$ grows rapidly with n. The conditional distribution is

$$\mathbb{P}(C_1 = c_1, \ldots, C_m = c_m | K(C) = k, N(C) = n) = \frac{n!}{|L_{n,k}^{(m)}|} \prod_{i=1}^m \frac{1}{c_i!}, \quad (1.8)$$

where $c \in \mathscr{S}_{n,k}$. The normalizing constant $|L_{n,k}^{(m)}|$ is a variant of the signless Lah number [10], which exemplifies an associated partial Bell polynomial introduced in the next chapter.

To calibrate the test, we must determine the proportion of $\mathscr{S}_{n,k}$ with chi-square value exceeding χ_*^2. Finding this proportion by enumerating all c is impractical. The solution can be approximated by sampling c from $\mathscr{S}_{n,k}$ using the Markov chain Monte Carlo (MCMC), whose stationary distribution is given by (1.8). Let $R = (r_{cc'})$ with

$$r_{cc'} = \mathbb{P}(C_{t+1} = c' | C_t = c)$$

be the transition matrix of a finite-state Markov chain that is irreducible, aperiodic, and symmetric. Then, the Markov chain with transition matrix $P = (p_{cc'})$, where

$$p_{cc'} = r_{cc'} \min\left\{1, \frac{\pi_{c'}}{\pi_c}\right\}, \quad c \neq c', \quad p_{cc} = 1 - \sum_{c' \neq c} p_{cc'}$$

has a unique stationary distribution π satisfying $\pi P = \pi$. Reversibility immediately follows from this observation, which provides the following Metropolis algorithm for sampling from the stationary distribution π.

Algorithm 1.1 (Metropolis et al., 1953 [11]) *A sequence of random variables from a distribution whose stationary distribution is the target distribution π is generated by the following algorithm:*

1. *Set $t = 0$ and pick an initial sample $c^{(0)}$.*
2. *Generate a candidate sample c' according to $\mathbb{P}(C' = c' | c^{(t)} = c) = r_{cc'}$.*
3. *Accept the candidate c' with probability*

$$\min\left\{1, \frac{\pi_{c'}}{\pi_c}\right\}$$

   and set $c^{(t+1)} = c'$. If rejected, set $c^{(t+1)} = c^{(t)}$.
4. *Increment t to $t + 1$ and return to Step 2.*

The Metropolis algorithm can draw samples from any probability distribution, provided that the acceptance ratio $\pi_{c'}/\pi_c$ of the target probability functions is available. This property renders the algorithm quite useful, because calculating the normalization constant is often difficult in practice. For a chain in the stationarity, the samples can be considered to be taken from the target distribution π. In the goodness-of-fit test discussed here, the proportion of samples with chi-squared values exceeding χ_*^2 estimates the significance probability (p-value).

To implement the Metropolis algorithm, we must construct a connected Markov chain over the support. A *Markov basis* is a finite set of moves that guarantees the connectivity of all support elements (the Markov basis will be formally defined in Chap. 5). The Markov chain can visit all elements by adding or subtracting moves constructed by the Markov basis. In Chap. 5, we will show that a Markov basis for the support $\mathscr{S}_{n,k}$ is

$$\{e_i + e_j - e_{i+1} - e_{j-1}; 1 \leq i < j \leq m; i + 2 \leq j\},$$

where e_i is the m-dimensional unit vector with 1 in the i-th component and 0 in all other components.

Example 1.2 [Poisson regression [9]] A chemical to control insects was successively sprayed onto equally infested plots. From plots 1 to 5, the concentration was 1, 2, 3, 4, and 5. The numbers of insects left alive on the plots were $(c_1, c_2, c_3, c_4, c_5) = (44, 25, 21, 19, 11)$. The data were subjected to the above-described goodness-of-fit test of the Poisson regression. The count vector gives $k = 120$, $n = 288$, and $\chi_*^2 = 1.686$. Following [9], the first 10, 000 samples were discarded and the significance probability was estimated from the next 90,000 samples. The estimated p-value was 0.023. This result will be validated in Example 5.2.

The drawbacks of MCMC algorithms are well recognized. Practical problems in implementation of MCMC samplers are discussed in [12]. Although the Markov chain eventually converges to the target distribution, the initial draws deviate from the target distribution. For example, when estimating the p-value above, we must consider not only the Monte Carlo error but also the error caused by departure from stationarity. Whether the chain is in the stationarity is not easily determined. Therefore, in actual implementations, the initial samples must be discarded in the so-called *burn-in* period. The appropriate burn-in period can be roughly estimated by investigating the convergence. This topic is called *mixing* of Markov chains. Another drawback is the autocorrelation among nearby samples in a chain. If we need independent samples, most of the samples must be discarded. Autocorrelation is commonly reduced by *thinning*, which retains every t-th sample in the chain over some long interval t. A huge number of steps is required for a reasonable estimate.

A sampler that draws independent samples directly from the target distribution would overcome the disadvantage of MCMC samplers. Chapter 3 introduces the A-hypergeometric distribution defined by Takayama et al. [13], which constitutes a class of discrete exponential families of distributions. Examples of A-hypergeometric distributions are the conditional EPPFs discussed in the previous subsection and the distribution given by (1.8). Chapter 5 presents a direct sampler for A-hypergeometric distributions proposed in [14]. Aided by properties of polytopes and the information geometry, Chaps. 3 and 5 discuss the maximum likelihood estimation of

A-hypergeometric distributions. The computations demand an evaluation of the normalization constants, which are the *A*-hypergeometric polynomials defined by Gel'fand et al. [15]. For small samples that may not be analyzed by asymptotic theory, the normalization constants must be evaluated numerically. Evaluation methods are discussed in Chap. 3.

References

1. Sibuya, M.: A random-clustering process. Ann. Inst. Statist. Math. **45**, 459–465 (1993)
2. Aldous, D.J.: Exchangeability and related topics. In: Ecole d'Été de Probabilités de Saint Flour, Lecture Notes in Mathematics, vol. 1117. Springer, Berlin (1985)
3. Pitman, J.: Combinatorial Stochastic Processes. Ecole d'Été de Probabilités de Saint Flour, Lecture Notes in Math. vol. 1875. Springer, Berlin (2006)
4. Kingman, J.F.C.: The representation of partition structures. J. Lond. Math. Soc. **18**, 374–380 (1978)
5. Lo, A.Y.: On a class of Bayesian nonparametric estimates: I. density estimates. Ann. Statist. **12**, 351–357 (1984)
6. Griffin, J., Holmes, C.: Computational issues arising in Bayesian nonparametric hierarchical models. In: Hjort, N.L., Holmes, C., Müller, P., Walker, S.G. (eds.) Bayesian Nonparametrics, pp. 208–222. Cambridge University Press, Cambridge (2010)
7. De Finetti, B.: La prévision: ses lois logiques, ses sources subjectives. Ann. Inst. H. Poincaré Probab. Statist. **7**, 1–68 (1937)
8. Ferguson, T.S.: A Bayesian analysis of some nonparametric problems. Ann. Statist. **1**, 209–230 (1973)
9. Diaconis, P., Eisenbud, B., Sturmfels, B.: Lattice walks and primary decomposition. In: Sagan, B.E., Stanley, R.P. (eds.) Mathematical Essays in Honor of Gian-Carlo Rota (Cambridge, MA, 1996). Progress in Mathematics, vol. 161, pp. 173–193. Birkhäuser, Boston (1998)
10. Charalambides, C.A.: Combinatorial Methods in Discrete Distributions. Wiley, New Jersey (2005)
11. Metropolis, N., Rosenbluth, A.W., Rosenbluth, M.N., Teller, A.H., Teller, E.: Equations of state calculations by fast computing machines. J. Chem. Phys. **21**, 1087–1092 (1953)
12. Gilks, W.R., Richardson, S., Spiegelhalter, D.J. (eds.): Markov Chain Monte Carlo in Practice. Chapman Hall, Boca Raton (1996)
13. Takayama, N., Kuriki, S., Takemura, A.: *A*-hypergeometric distributions and Newton polytopes. Adv. in Appl. Math. **99**, 109–133 (2018)
14. Mano, S.: Partition structure and the *A*-hypergeometric distribution associated with the rational normal curve. Electron. J. Stat. **11**, 4452–4487 (2017)
15. Gel'fand, I.M., Kapranov, M.M., Zelevinsky, A.V.: Generalized Euler integrals and *A*-hypergeometric functions. Adv. in Math. **84**, 255–271 (1990)

Chapter 2
Measures on Partitions

Abstract After brief introduction of the multiplicative measure, defined as a family of measures on integer partitions, which include typical combinatorial structures, this chapter introduces the exponential structure, which plays important roles in statistical inference. It then introduces the Gibbs partition, a generalization of the exponential structure. The generalization is achieved by systematic use of partial Bell polynomials. Gibbs partitions characterize prior processes in Bayesian nonparametrics, and appear as statistical models of diversity in count data. The Ewens sampling formula and the Pitman partition are well-known examples of Gibbs partitions. Finally, this section discusses the asymptotic behaviors of extremes of the sizes of parts in Gibbs partitions. Some of the results are derived by simple analytic approaches.

Keywords Analytic combinatorics · Asymptotics · Ewens sampling formula
Exponential structure · Extreme · Gibbs partition · Multiplicative measure
Partial bell polynomial · Pitman partition

2.1 Multiplicative Measures

Let the set of integer partitions of n be denoted by $\mathscr{P}_n := \{\lambda; \lambda \vdash n\}$, and the subset of length k by

$$\mathscr{P}_{n,k} := \{\lambda; \lambda \vdash n, l(\lambda) = k\},$$

where $\mathscr{P}_n = \cup_k \mathscr{P}_{n,k}$.

Example 2.1 The number 4 is partitioned as

$$\mathscr{P}_4 = \{(4), (3, 1), (2, 2), (2, 1, 1), (1, 1, 1, 1)\}.$$

For each length, we have $\mathscr{P}_{4,1} = \{(4)\}$, $\mathscr{P}_{4,2} = \{(3, 1), (2, 2)\}$, $\mathscr{P}_{4,3} = \{(2, 1, 1)\}$, $\mathscr{P}_{4,4} = \{(1, 1, 1, 1)\}$.

Integer partitions can be represented in a *Young tableau*, a collection of boxes arranged in left-justified boxes, with the row length in non-increasing order. An integer partition $\lambda \in \mathscr{P}_n$ is uniquely determined by listing the number of boxes in each

© The Author(s) 2018
S. Mano, *Partitions, Hypergeometric Systems, and Dirichlet Processes in Statistics*,
JSS Research Series in Statistics, https://doi.org/10.1007/978-4-431-55888-0_2

Fig. 2.1 Young tableau of
the partition $(4, 2, 1) \in \mathscr{P}_{7,3}$

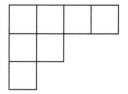

row. An example is given in Fig. 2.1. The size index $c_i(\lambda) := \#\{j; \lambda_j = i\}$ is the
number of rows with length i. In the present example, we have $c_4 = c_2 = c_1 = 1$.

Stochastic processes on Young tableaux lie at intersection of probability theory
and representation theory, and have been extensively studied. Some pioneering works
are [1, 2]. Recent developments can be found in [3, 4] and references therein. As
is well known, partitioning integer into multisets and powersets (defined below),
corresponds to partitioning a given energy into bosons and fermions, respectively.
Therefore, these combinatorial structures have been important models for analyzing
numbers of states in systems in statistical mechanics [5]. Analyses of the modular
forms appearing in generating functions are also central to string and field theories.
For the details, refer to [6, 7] and references therein. Integer partitions appear as
statistical models of diversity in count data. An example is application to statistical
disclosure control. A comprehensive survey on the subject in Japanese is [8].

Vershik [5] discussed limit theorems for random functions induced by a probability measure on Young tableaux. The right continuous step function

$$\varphi_\lambda(t) := \sum_{i \geq t} c_i(\lambda), \qquad 0 \leq t < \infty,$$

satisfying

$$\int_0^\infty \varphi_\lambda(t) dt = n,$$

encodes the shape of a Young tableau. Vershik defined a family of discrete probability
measures on $\mathscr{P}_n$, and called it the *multiplicative measure*. For a sequence of positive
real-number valued functions $s_i(c)$, $i \in \mathbb{N}$ with $s_i(0) = 1$, let

$$F(\lambda) = \prod_{i \geq 1} s_i(c_i(\lambda)). \tag{2.1}$$

The multiplicative measure is defined as

$$\mu_n(\lambda) := Z_n^{-1} F(\lambda), \qquad \lambda \in \mathscr{P}_n, \qquad Z_n := \sum_{\lambda \vdash n} F(\lambda). \tag{2.2}$$

The multiplicative measure is intuitively described as follows. For a combinatorial
structure determined by partitions, $F(\lambda)$ is proportional to the number of possible

structures with the partition λ, where all possible structures can occur with equal probability. The ordinary generating functions of Z_n and $(s_i(c))$ respectively given by

$$\mathscr{F}(x) = \sum_{n \geq 0} Z_n x^n, \qquad \mathscr{F}_i(y) = \sum_{c \geq 0} s_i(c) y^c,$$

are related through the *multiplicativity*:

$$\mathscr{F}(x) = \sum_{n \geq 0} \sum_{\{c: \sum ic_i = n\}} \prod_{i \geq 1} s_i(c_i) x^n = \prod_{i \geq 1} \sum_{c \geq 0} s_i(c) x^{ic} = \prod_{i \geq 1} \mathscr{F}_i(x^i). \qquad (2.3)$$

For a given length $l(\lambda) = k$, the conditional probability measure is given by

$$\mu_{n,k}(\lambda) = Z_{n,k}^{-1} F(\lambda), \qquad \lambda \in \mathscr{P}_{n,k}, \qquad Z_{n,k} := \sum_{\lambda \in \mathscr{P}_{n,k}} F(\lambda), \qquad (2.4)$$

where $Z_n = \sum_{k=1}^{n} Z_{n,k}$.

Let us focus on three classes of functions for $(s_i(c))$. These classes, which yield typical combinatorial structures with simple probabilistic interpretations, are the negative binomial coefficients, the binomial coefficients, and the coefficients appearing in the Poisson distribution. They are respectively given by

$$\binom{w_i + c - 1}{c}, \qquad \binom{w_i}{c}, \qquad \left(\frac{w_i}{i!}\right)^c \frac{1}{c!}, \qquad i \in \mathbb{N},$$

for some sequence of positive integers (w_i). The combinatorial structures given by the negative binomial coefficients, the binomial coefficients, and the coefficient in the Poisson distribution are called *multisets*, *powersets* (or *selections*), and *exponential structures* (or *assemblies*), respectively. Arratia et al. [9] defined the *conditioning relation*, the property under which the joint distribution of the size indices satisfies

$$(C_1, C_2, ..., C_n) \stackrel{d}{=} (X_1, X_2, ..., X_n | X_1 + 2 \cdot X_2 + \cdots + n \cdot X_n = n) \qquad (2.5)$$

for a sequence of independent random variables $(X_1, ..., X_n)$. The above three coefficients come from the distributions followed by $(X_1, ..., X_n)$. Further examples of these three combinatorial structures, beyond the limited examples given here, are given in Chap. 2 of [9] and references therein. Chapter 5 of [10] is devoted to exponential structures. Chapters 1 and 2 of [11] provide systematic accounts of various aspects of combinatorial structures using symbolic methods.

Multisets are combinatorial structures wherein each of c_i parts of size i takes one of the w_i possible structures. The generating function of $(s_i(c))$ is

$$\mathscr{F}_i(y) = \sum_{c \geq 0} \binom{w_i + c - 1}{c} y^c = \sum_{c \geq 0} \binom{-w_i}{c} (-y)^c = (1 - y)^{-w_i}.$$

By straightforward observation, one finds that the conditioning relation (2.5) is satisfied by independent random variables $X_i \sim \text{NegBin}(w_i, \zeta^i)$, $i \in \mathbb{N}$, for some $\zeta \in (0, 1)$.

Example 2.2 (*Integer partitions*) The integer partition is a multiset with $w_i = 1$. As $F(\lambda) = 1$, both $\mu_n(\lambda)$ and $\mu_{n,k}(\lambda)$ are uniform distributions on the integer partitions in each support. For $n = 4$ in Example 2.1, we have $Z_4 = 5$, so each partition occurs with probability $1/5$. For the subset with length $l(\lambda) = 2$, $Z_{4,2} = 2$, so each partition occurs with probability $1/2$. The number of distinct integer partitions Z_n is called the partition number. The well-known Hardy–Ramanujan formula gives

$$Z_n \sim \frac{1}{4n\sqrt{3}} \exp\left(2\pi\sqrt{\frac{n}{6}}\right), \qquad n \to \infty.$$

This result can be obtained by evaluating the residue around the singularity at $x = 1$ of the generating function $\mathscr{F}(x) = \prod_{i \geq 1}(1 - x^i)^{-1}$.

Powersets are combinatorial structures like multisets, but are partitioned as subsets rather than multisubsets. In powersets, each of c_i parts of size i takes one of the w_i structures, and must be distinct from all other parts. The generating function of $(s_i(c))$ is

$$\mathscr{F}_i(y) = \sum_{c \geq 0} \binom{w_i}{c} y^c = (1 + y)^{w_i}.$$

The conditioning relation (2.5) is satisfied by independent random variables $X_i \sim \text{Bin}(w_i, \zeta^i/(1 + x^i))$, $i \in \mathbb{N}$, for some $\zeta > 0$.

Example 2.3 (*Integer partitions*) An integer partition with distinct integers is a powerset with $w_i = 1$. As $F(\lambda) = 1$, both $\mu_n(\lambda)$ and $\mu_{n,k}(\lambda)$ are uniform distributions on the integer partitions in each support. In contrast to the multiset in Example 2.2, the support does not consist of all partitions. For $n = 4$ in Example 2.1, the support is $\{(4), (3, 1)\}$, and each partition occurs with probability $1/2$.

Finally, let us visit exponential structures. The objects are unlabeled in multisets and powersets, but are labeled in exponential structures. A typical example of exponential structures is the partition of a finite set. A finite-set partition induces an integer partition while ignoring the labels, as will be shown in Example 2.4. Labeled objects are better discussed by exponential generating functions than ordinary generating functions. Chapter 2 of [11] discusses a combinatorial structure called *sets*, an unlabeled version of multisets and powersets. An exponential structure is the composition of a set and a sequence of positive integers (w_i). Exponential structures play important roles in statistical applications, as each observation can be usually labeled, and the Poisson distribution is the standard model of count data.

Definition 2.1 Assume that a set $[n]$ is partitioned into parts. For each part of size i, one of w_i possible structures is chosen independently. The resulting structure is called an *exponential structure*.

Let us count the number of possible instances of size n for a given size index c. The number of ways is

$$n! \prod_{i=1}^{n} \left(\frac{1}{i!}\right)^{c_i} \frac{1}{c_i!}.$$

For each part of size i, there are w_i possible structures. Permutation with repetition then yields

$$n! \prod_{i=1}^{n} \left(\frac{w_i}{i!}\right)^{c_i} \frac{1}{c_i!}.$$

Let us set

$$F(\lambda) = n! \prod_{i \geq 1} \left(\frac{w_i}{i!}\right)^{c_i(\lambda)} \frac{1}{c_i(\lambda)!}.$$

Here, the expression (2.1) is multiplied by $n!$ because we are dealing with labeled structures. The generating function of $(s_i(c)), i \in \mathbb{N}$ is

$$\mathscr{F}_i(y) = \sum_{c \geq 0} \left(\frac{w_i}{i!} y\right)^c \frac{1}{c!} = \exp\left(\frac{w_i}{i!} y\right).$$

Using the exponential generating function of the sequence of positive real numbers $w_i, i \in \mathbb{N}$:

$$\mathscr{W}(x) = \sum_{i \geq 1} \frac{w_i}{i!} x^i, \tag{2.6}$$

and the relationship between the generating functions (2.3), we obtain the exponential generating function of $Z_n = \sum_{\lambda \vdash n} F(\lambda)$:

$$\mathscr{F}(x) = e^{\mathscr{W}(x)}. \tag{2.7}$$

This formula, called the *exponential formula*, is a characterizing property of exponential structures [10]. The conditioning relation (2.5) is satisfied by independent random variables $X_i \sim \text{Po}(w_i \zeta^i / i!), i \in \mathbb{N}$, for some $\zeta > 0$.

Example 2.4 (Set partitions) The partition of a finite set, or *set partition*, is an exponential structure with $w_i = 1$. The number of possible instances of size n in a set that is called the n-th Bell number B_n. Thus, $Z_n = B_n$ and the exponential generating function is $\mathscr{F}(x) = \exp(e^x - 1)$. Moreover, $Z_{n,k}$ defines the number of ways of partitioning a set of n objects into k non-empty subsets. The number is called the Stirling number of the second kind, denoted by $S(n, k)$. As $B_n = \sum_{k=1}^{n} S(n, k)$, the multiplicative measure (2.2) becomes

$$\mu_n(\lambda) = \frac{n!}{B_n} \prod_{i=1}^{n} \left(\frac{1}{i!}\right)^{c_i} \frac{1}{c_i!}, \qquad \lambda \in \mathscr{P}_n.$$

For a given length $l(\lambda) = k$, the conditional probability measure (2.4) is

$$\mu_{n,k}(\lambda) = \frac{n!}{S(n,k)} \prod_{i=1}^{n} \left(\frac{1}{i!}\right)^{c_i} \frac{1}{c_i!}, \qquad \lambda \in \mathscr{P}_{n,k}.$$

Consider the integer partitions of $n = 4$ in Example 2.1. The subset of integer partitions of length $l(\lambda) = 2$ is $\mathscr{P}_{4,2} = \{(3, 1), (2, 2)\}$. Each of the integer partitions determines a subset of partitions of finite sets with identified cardinalities. For $(3, 1)$ and $(2, 2)$, the instances are

$$\{\{\{1, 2, 3\}, \{4\}\}, \{\{2, 3, 4\}, \{1\}\}, \{\{3, 4, 1\}, \{2\}\}, \{\{4, 1, 2\}, \{3\}\}\}$$

and

$$\{\{\{1, 2\}, \{3, 4\}\}, \{\{1, 3\}, \{2, 4\}\}, \{\{1, 4\}, \{2, 3\}\}\},$$

respectively. As $S(4, 2) = 7$ and $B_4 = 15$, we have

$$\mu_{4,2}((3, 1)) = \frac{4}{7}, \qquad \mu_{4,2}((2, 2)) = \frac{3}{7},$$

and

$$\mu_4((3, 1)) = \frac{4}{15}, \qquad \mu_4((2, 2)) = \frac{3}{15}.$$

Remark 2.1 In Example 2.4, probability measures on the integer partitions were induced by the equivalence relations under the actions of the symmetric group on the uniform distribution on set partitions.

Example 2.5 (Permutations) A permutation may be viewed as a set of cyclic permutations. Figure 2.2 shows the decomposition of the permutation

$$\begin{pmatrix} 1\ 2\ 3\ 4\ 5\ 6 \\ 2\ 4\ 5\ 1\ 3\ 6 \end{pmatrix} = (1, 2, 4)(3, 5)(6),$$

where the parentheses on the right hand side denote cycles. There are $(i - 1)!$ ways to place i elements in a cycle. The decomposition of a permutation is an exponential structure with $w_i = (i - 1)!$. As the number of permutations of $[n]$ is $n!$, we have $Z_n = n!$. The exponential generating function is $\mathscr{F}(x) = (1 - x)^{-1}$, and $Z_{n,k}$ is the number of ways of decomposing a permutation of $[n]$ into k cycles, called the signless Stirling number of the first kind, denoted by $|s(n, k)|$. The multiplicative measure (2.2) becomes

$$\mu_n(\lambda) = \prod_{i=1}^{n} \left(\frac{1}{i}\right)^{c_i} \frac{1}{c_i!}, \qquad \lambda \in \mathscr{P}_n.$$

Fig. 2.2 A permutation
represented by cycles

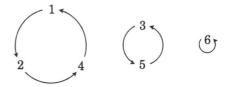

For a given length $l(\lambda) = k$, the conditional probability measure (2.4) is

$$\mu_{n,k}(\lambda) = \frac{n!}{|s(n,k)|} \prod_{i=1}^{n} \left(\frac{1}{i}\right)^{c_i} \frac{1}{c_i!}, \qquad \lambda \in \mathscr{P}_{n,k}.$$

Each integer partition determines the subsets of permutations with identified cycle lengths. For $(3, 1)$ and $(2, 2)$, we have $\mathscr{P}_{4,2} = \{(3, 1), (2, 2)\}$ with instances

$\{(1, 2, 3)(4), (1, 3, 2)(4), (2, 3, 4)(1), (2, 4, 3)(1), (3, 4, 1)(2), (3, 1, 4)(2),$
$(4, 1, 2)(3), (4, 2, 1)(3)\}$

and

$$\{(1, 2)(3, 4), (1, 3)(2, 4), (1, 4)(2, 3)\}.$$

As $|s(4, 2)| = 11$ and $4! = 24$, we have

$$\mu_{4,2}((3, 1)) = \frac{8}{11}, \qquad \mu_{4,2}((2, 2)) = \frac{3}{11},$$

and

$$\mu_4((3, 1)) = \frac{1}{3}, \qquad \mu_4((2, 2)) = \frac{1}{8}.$$

Remark 2.2 For permutations, the exponential formula (2.7) takes the form

$$\mathscr{F}(x) = e^{\mathscr{W}(x)} = \exp(-\log(1 - x)).$$

A class of combinatorial structures whose members take this exponential-of-logarithm form as the generating function is called *exp-log class*. Flajolet and Soria [12] introduced this class and discussed the limit laws of its length of partitions.

Example 2.6 (*Forests of labeled rooted trees*) A rooted tree is a connected directed acyclic graph with one distinguished vertex. To construct a forest of rooted trees, the $[n]$ labeled vertices are partitioned into subsets, and a rooted tree is made from all vertices in each subset. The number of non-plane labeled rooted trees that can be constructed from i vertices is given by Cayley's formula, i^{i-1}. Forests of rooted trees are exponential structures with $w_i = i^{i-1}$. Considering each vertex adjacent to 0 is the root of a tree, there exists a bijection between a single unrooted tree on vertices

$\{0, 1, \ldots, n\}$ and a forest of rooted trees on vertices $[n]$. Moreover, the number of unrooted trees that can be constructed from $(n + 1)$ vertices is $(n + 1)$ times the number of rooted trees that can be constructed from $(n + 1)$ vertices. Therefore, the number of forests of labeled rooted trees is $Z_n = (n + 1)^{n-1}$. The exponential formula (2.7) yields a non-trivial identity:

$$\mathscr{F}(x) = \sum_{n \geq 0} \frac{\{(n + 1)x\}^n}{(n + 1)!} = \exp\left(\sum_{i \geq 1} \frac{i^{i-1}}{i!} x^i\right) = e^{-W_0(-x)}, \tag{2.8}$$

where $W_0(x)$ defines the principal branch of the Lambert W function (product logarithm) [13], which is the branch of inverse relation of the function $f(z) = ze^z$ with $z \geq -1$. The combinatorics is further discussed in Sect. 2.5.1 of [11]. Each integer partition determines forests with identified sizes of trees. For $\{(3, 1), (2, 2)\}$, the number of possible forests are 36 for $(3, 1)$ and 12 for $(2, 2)$, respectively. Therefore, we have

$$\mu_{4,2}((3, 1)) = \frac{3}{4}, \qquad \mu_{4,2}((2, 2)) = \frac{1}{4}$$

and

$$\mu_4((3, 1)) = \frac{36}{125}, \qquad \mu_4((2, 2)) = \frac{12}{125}.$$

An instance of size $(3, 1)$ is shown in Fig. 2.3.

A multiplicative measure can be obtained from a given multiplicative measure by a simple operation called *tilting*. Tilting is achieved by substituting

$$F(\lambda) \mapsto \theta^{l(\lambda)} F(\lambda), \qquad \theta > 0, \tag{2.9}$$

into the expression (2.1). Tilting weights the probability mass function of the multiplicative measure (2.2) by the length. In particular, for exponential structures, tilting (2.9) is equivalent to the mapping

$$w_i \mapsto \theta w_i, \qquad i \in \mathbb{N}, \qquad \theta > 0. \tag{2.10}$$

Note that the exponential structure is invariant under this mapping

$$w_i \mapsto \phi^i w_i, \qquad i \in \mathbb{N}, \qquad \phi > 0. \tag{2.11}$$

Fig. 2.3 A forest of labeled rooted trees of size $(3, 1)$

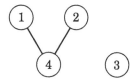

Remark 2.3 Unlike the marginal probability measure, the conditional probability measure for a given length (2.4) is invariant under tilting. This shows that the length is the sufficient statistic for the tilting parameter θ.

Example 2.7 (Mappings) Consider all mappings form $[n]$ to itself. The number of mapping is $Z_n = n^n$. A mapping f corresponds to a directed graph with vertices $[n]$ and edges $(i, f(i)), i \in [n]$, where the outdegree of every vertex is one. The graph is decomposed into connected subgraphs. Katz [14] showed that i labeled vertices can be arranged into a connected graph in $f_i = (i - 1)! \sum_{j=0}^{i-1} i^j / j!$ ways. By the central limit theorem, we have

$$\frac{f_i}{i!} = \frac{e^i}{i} \mathbb{P}(Po(i) \leq i - 1) \sim \frac{e^i}{2i}, \quad i \to \infty,$$

where $Po(i)$ is the Poisson random variable of mean i. From (2.10) and (2.11), we find that the exponential structure given by the mapping decomposition is asymptotically equivalent to that of decomposition of permutation into cycles in Example 2.5 with tilting (2.9) with $\theta = 1/2$. Each integer partition determines subsets of mappings with identified size of connected subgraphs. For $\{(3, 1), (2, 2)\}$, the numbers of possible mappings are 68 for (3, 1) and 27 for (2, 2), respectively. Therefore, we have

$$\mu_4((3, 1)) = \frac{68}{4^4} = \frac{17}{64} = 0.266..., \quad \mu_{4,2}((3, 1)) = \frac{68}{95} = 0.716...$$

Meanwhile, the decomposition of permutation into cycles with tilting of $1/2$ gives

$$\mu_4((3, 1)) = \frac{32}{105} = 0.305..., \quad \mu_{4,2}((3, 1)) = \frac{8}{11} = 0.727...$$

As expected from Remark 2.3, the conditional probability measure determined from the number of connected subgraphs approximates that obtained by decomposing a random permutation into cycles with tilting of $1/2$. An instance of mapping of size (3, 1) is shown in Fig. 2.4.

Let us collect some formulae of multiplicative measures (2.2) induced by decomposing the permutation into cycles, as shown in Example 2.5 with tilting (2.9). It is straightforward to see that $Z_{n,k} = \theta^k |s(n, k)|$. Using the exponential formula (2.7) and the negative binomial theorem, we obtain

Fig. 2.4 A mapping of size (3, 1)

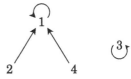

$$Z_n = \left[\frac{x^n}{n!}\right] \mathscr{F}(x) = \left[\frac{x^n}{n!}\right] (1 - x)^{-\theta} = (\theta)_n,$$

where $[x^n/n!]f(x)$ represents the coefficient of $x^n/n!$ in expansion of $f(x)$ as a power series in x. The multiplicative measure is

$$\mu_n(\lambda) = \frac{n!}{(\theta)_n} \prod_{i=1}^{n} \left(\frac{\theta}{i}\right)^{c_i} \frac{1}{c_i!}, \qquad \lambda \in \mathscr{P}_n. \tag{2.12}$$

This expression is known as the Ewens sampling formula [15], appeared in sampling from genetic diversity (see Appendix A.2). This sampling formula was independently derived by Antoniak [16] as the sampling distribution of the Dirichlet process (see Chap. 4). The background is discussed in [17–19]. The conditional probability measure given the length $l(\lambda) = k$ (2.4) is

$$\mu_{n,k}(\lambda) = \frac{n!}{|s(n,k)|} \prod_{i=1}^{n} \left(\frac{1}{i}\right)^{c_i} \frac{1}{c_i!}, \qquad \lambda \in \mathscr{P}_{n,k}. \tag{2.13}$$

2.2 Exponential Structures and Partial Bell Polynomials

Thus far, we have discussed multiplicative measures induced by exponential structures. In this section, we generalize these measures to a family of discrete probability measures on integer partitions $\mathscr{P}_n$. The length of partition is recognized as an important statistic. For example, in the Bayesian mixture modeling discussed in Sect. 1.2.1, the length is the number of clusters. Although the distribution of the length must be flexible, the multiplicative measures admit only weights by the power of length, namely, the tilting factor (2.9). The multiplicative measures induced by exponential structures can be generalized using relatives of Bell polynomials [20]. For an extensive survey on this subject, see Chap. 1 of [21].

Consider a sequence of positive real numbers $(v_{n,k})$, $k \in [n]$, $n \in \mathbb{N}$, and begin with

$$F(\lambda) = v_{n,l(\lambda)} n! \prod_{i \geq 1} \left(\frac{w_i}{i!}\right)^{c_i} \frac{1}{c_i!}.$$

Moreover, let (w_i), $i \in \mathbb{N}$ be a sequence of positive real numbers (not restricted to be an integer sequence). Identically to the case for the multiplicative measure in (2.4), the conditional probability measure given length $l(\lambda) = k$ is

$$\mu_{n,k}(\lambda) = \frac{n!}{B_{n,k}(w)} \prod_{i=1}^{n} \left(\frac{w_i}{i!}\right)^{c_i} \frac{1}{c_i!}, \qquad \lambda \in \mathscr{P}_{n,k}, \tag{2.14}$$

Table 2.1 The partial Bell polynomials.

$n \backslash k$	1	2	3	4	5
1	w_1				
2	w_2	w_1^2			
3	w_3	$3w_1w_2$	w_1^3		
4	w_4	$4w_1w_3 + 3w_2^2$	$6w_1^2w_2$	w_1^4	
5	w_5	$5w_1w_4 + 10w_2w_3$	$10w_1^2w_3 + 15w_1w_2^2$	$10w_1^3w_2$	w_1^5

where

$$Z_{n,k} = v_{n,k}B_{n,k}(w), \qquad B_{n,k}(w) := n! \sum_{\lambda \vdash n, l(\lambda)=k} \prod_{i=1}^{n} \left(\frac{w_i}{i!}\right)^{c_i} \frac{1}{c_i!}. \qquad (2.15)$$

The probability function (2.14) is a discrete exponential family with natural parameters $(\log(w_1/1!), ..., \log(w_n/n!))$ and sufficient statistics $(c_1, ..., c_n)$. The normalization constant $B_{n,k}(w)$ is called the *partial Bell polynomial*, which defines the number of exponential structures consisting of n elements with k parts. The partial Bell polynomials are polynomials of the variables w_i, $i \in \mathbb{N}$. Some explicit expressions for small n are given in Table 2.1. As the convention, we set $B_{n,0}(w) = \delta_{n,0}$. The *Bell polynomials* are then given by $B_n(w) := \sum_{k=1}^{n} B_{n,k}(w)$. Especially, the n-th Bell number B_n is $B_n(1.)$, $1. = (1, 1, ...)$. Chapter 3 of Comtet's book [22] explains formal series. Partial Bell polynomials play fundamental roles in the discussion and various results on partial Bell polynomials can be found in the chapter. The multiplicative measure (2.2) induced by the exponential structure now becomes

$$\mu_n(\lambda) = \frac{n!}{B_n(w)} \prod_{i=1}^{n} \left(\frac{w_i}{i!}\right)^{c_i} \frac{1}{c_i!}, \qquad \lambda \in \mathscr{P}_n. \qquad (2.16)$$

The *weighted Bell polynomials* are given by

$$B_n(v, w) := \sum_{k=1}^{n} v_{n,k}B_{n,k}(w)$$

The partial Bell polynomials can be expressed by an exponential generating function of the sequence w_i, $i \geq 1$ (2.6):

$$\left[\frac{x^n}{n!}\right]\frac{\{\mathscr{W}(x)\}^k}{k!} = \left[\frac{x^n}{n!}\right]\frac{1}{k!}\left(\sum_{i \geq 1}\frac{w_i}{i!}x^i\right)^k = \sum_{n_1+\cdots+n_k=n} \prod_{i=1}^{k}\frac{w_{n_i}}{n_i!} = B_{n,k}(w).$$

An equivalent exponential generating function of the partial Bell polynomials is

$$\sum_{n \geq 1} B_{n,k}(w) \frac{x^n}{n!} = \frac{\{\mathscr{W}(x)\}^k}{k!}. \tag{2.17}$$

Suppose that the sequence $(v_{n,k})$ does not depend on n, say (v_k). By using the exponential generating function

$$\mathscr{V}(x) = \sum_{k \geq 1} v_k \frac{x^k}{k!},$$

we have *Faà di Bruno's formula*

$$B_n(v, w) = \left[\frac{x^n}{n!}\right] \mathscr{V}(\mathscr{W}(x)). \tag{2.18}$$

Especially, for a tilting (2.9) with $(v_k) = (\theta^k)$, we have $B_n(\theta^{\cdot}, w_{\cdot}) = B_n(\theta w_{\cdot})$.

Remark 2.4 Faà di Bruno's formula gives successive derivatives of composition of functions:

$$\{\mathscr{V}(\mathscr{W}(x))\}^{(n)} = B_n(\mathscr{V}^{(\cdot)}(\mathscr{W}(x)), \mathscr{W}^{(\cdot)}(x)),$$

where the superscript (n) means the n-th derivative. The formula (2.18) provides the value at $x = 0$. Further discussion is given in Sect. 3.4 of [22].

Some examples are demonstrated below.

Example 2.8 (Stirling number of the second kind) In Example 2.4, $w_i = 1$ and $\mathscr{W}(x) = e^x - 1$. Now set $v_k = 1$, $Z_{n,k} = S(n, k) = B_{n,k}(1_{\cdot})$, and let $v_{n,k} = [\theta]_k$, $\mathscr{V}_n(x) = (1 + x)^{\theta} - 1$, where the falling factorial $[x]_i := x(x - 1) \cdots (x - i + 1)$ is used. From faà di Bruno's formula (2.18), we obtain the factorial generating function

$$B_n([\theta]_{\cdot}, 1_{\cdot}) = \sum_{k=1}^{n} [\theta]_k S(n, k) = \theta^n.$$

Example 2.9 (Signless Stirling number of the first kind) In Example 2.5, $w_i = (i - 1)!$ and $\mathscr{W}(x) = -\log(1 - x)$. Now set $v_k = 1$, $Z_{n,k} = |s(n, k)| = B_{n,k}((\cdot - 1)!)$, and let $v_k = \theta^k$, $\mathscr{V}_n(x) = e^{\theta x} - 1$. From Faà di Bruno's formula (2.18), we obtain the ordinary generating function

$$B_n(\theta^{\cdot}, (\cdot - 1)!) = \sum_{k=1}^{n} \theta^k |s(n, k)| = (\theta)_n. \tag{2.19}$$

Example 2.10 (*Tilted forests of labeled rooted trees*) In Example 2.6, $w_i = i^{i-1}$ and $\mathcal{W}(x) = -W_0(-x)$. Although the partial Bell polynomial $Z_{n,k} = B_{n,k}(\cdot^{-1})$ has no specific name, a weighted Bell polynomial has a closed form. Setting $v_k = \theta^k$, $\mathcal{V}(x) = e^{\theta x} - 1$, Faà di Bruno's formula (2.18) yields the ordinary generating function

$$B_n(\theta^{\cdot}, \cdot^{-1}) = \sum_{k=1}^{n} \theta^k B_{n,k}(\cdot^{-1}) = \left[\frac{x^n}{n!}\right] e^{-\theta W_0(-x)} = \theta(\theta + n)^{n-1}, \qquad (2.20)$$

where the last equality follows by the *Lagrange inversion formula* (see Sect. A6 of [11]). Let $\phi(u) = \sum_{k \geq 0} \phi_k u^k$ with $\phi_0 \neq 0$. The equation $y = x\phi(y)$ then admits a unique solution with coefficients given by

$$y(x) = \sum_{i \geq 1} y_i x^i, \qquad y_i = \frac{1}{i}[u^{i-1}](\phi(u))^i.$$

The series expression of the Lambert W function in (2.8) directly follows from the Lagrange inversion. As $-W_0(-x) = x \exp(-W_0(-x))$, we identify $\phi(u) = e^u$ and immediately obtain as

$$y(x) = -W_0(-x) = \sum_{i \geq 1} \frac{i^{i-1}}{i!} x^i.$$

The Lagrange-Bürmann inversion formula for fractional powers is given by

$$[x^n]\left(\frac{y(x)}{x}\right)^{\theta} = \frac{\theta}{n + \theta}[u^n](\phi(u))^{n+\theta}, \qquad \theta \in \mathbb{C}.$$

Setting $y(x)/x = -W_0(-x)$ immediately yields (2.20). The partial Bell polynomial also has a closed form:

$$B_{n,k}(\cdot^{-1}) = \binom{n-1}{k-1} n^{n-k}.$$

Moreover, $B_n(\theta \cdot^{-1}) = \theta(\theta + n)^{n-1}$.

Example 2.11 (*Macdonald symmetric functions*) This example is related to a family of symmetric functions known as the Macdonald symmetric functions [23]. The properties of the Macdonald symmetric functions used in this monograph are briefly summarized in Sect. A.1. Setting

$$w_i = \frac{t^i - 1}{q^i - 1}(i - 1)!,$$

where $q, t > 1$ or $q, t < 1$, the identity (A.6) implies that

$$B_n\left(\frac{t^\cdot - 1}{q^\cdot - 1}(\cdot - 1)!\right) = \frac{(t; q)_n}{(q; q)_n} n!, \qquad (x; y)_n := \prod_{i=0}^{n-1}(1 - xy^i).$$

Taking $t = q^\theta$ with $q \to 1$, this expression reduces to (2.19). In this sense, the exponential structure determined by the sequence $w_i = (i - 1)!(q^{i\theta} - 1)/(q^i - 1)$ is a "q-deform" of the tilted decomposition of permutations into cycles.

Example 2.12 (*Hermite polynomials*) This monograph considers positive real sequences for (v_k) and (w_i), since probability measures are constructed from such sequences. However, Okano et al. [24] showed that

$$B_n((x, -1, 0, 0, ...)) = h_n(x),$$

where $h_n(x)$ are the Hermite polynomials

$$h_n(x) = (-1)^n e^{x^2/x} \frac{d^n}{dx^n} e^{-x^2/2}, \qquad n \in \mathbb{N}_0 := \{0, 1, 2, ...\}.$$

Applying Faà di Burno's formula, they also proved properties of the Hermite polynomials. The Hermite polynomials with negative indices will appear in Example 4.12.

After normalization, the marginal probability measure on $\mathscr{P}_n$ is given as

$$\mu_n(\lambda) = \frac{v_{n,l(\lambda)} n!}{B_n(v, w)} \prod_{i=1}^{n} \left(\frac{w_i}{i!}\right)^{c_i} \frac{1}{c_i!}, \qquad \lambda \in \mathscr{P}_n. \tag{2.21}$$

When $v_{n,k} = 1$, the probability function (2.21) reduces to a multiplicative measure induced by exponential structures. The tilted measure is then obtained by taking $v_{n,k} = \theta^k$. These special probability functions are discrete exponential family. However, in general, the probability function (2.21) is not an exponential family because it depends on the length $l(\lambda)$ via $v_{n,l(\lambda)}$. In other words, the length $l(\lambda)$ is the sufficient statistic for the sequence $(v_{n,k})$, and the conditional probability measure (2.14) is an exponential family. In this monograph, we call the probability measure (2.21) the *Gibbs partition*. The distribution of the length of partition is

$$\mathbb{P}(l(\Lambda) = k) = \frac{v_{n,k} B_{n,k}(w)}{B_n(v, w)}. \tag{2.22}$$

Remark 2.5 The Gibbs partition is not consistently defined in the literature. Infinite exchangeability (see Sect. 1.1) is not assumed in this monograph, but is usually assumed in Bayesian statistics literature. The definition in Pitman's book [21] does not assume infinite exchangeability, but assumes that $(v_{n,k})$ is independent of n. In this monograph, we assume neither infinite exchangeability nor the form of $(v_{n,k})$.

If the sequence $(v_{n,k})$ is independent of n, the above construction of the Gibbs partition has a simple probabilistic interpretation, and the model is specifically called *Kolchin's model* [25, 26]. Let $(X_1, X_2, ...)$ be a sequence of i.i.d. random variables and K be a random variable independent of $(X_1, X_2, \ldots)$. Suppose X_i and K follow the power series distributions

$$\mathbb{P}(X_i = x) = \frac{w_x}{\mathscr{W}(\xi)} \frac{\xi^x}{x!}, \qquad x \in \mathbb{N},$$

$$\mathbb{P}(K = k) = \frac{v_k}{\mathscr{V}_0(\zeta)} \frac{\zeta^k}{k!}, \qquad k \in \mathbb{N}_0,$$

where $\zeta = \mathscr{W}(\xi)$ and $\mathscr{V}_0(\zeta)$ is the exponential generating function of (v_k), $k \in \mathbb{N}_0$. From the composition of probability generating functions of X_i and K, the distribution of $S_K := X_1 + \cdots + X_K$ is obtained as

$$\mathbb{P}(S_K = n) = \frac{\xi^n}{n!} \frac{B_n(v, w)}{\mathscr{V}_0(\mathscr{W}(\xi))},$$

where we set $S_0 = 0$. If $v_k = 1$, $\mathscr{V}_0(\zeta) = e^\zeta$ and S_K follow a compound Poisson distribution. Meanwhile, the multiplicative measure reduces to that of an exponential structure (2.16).

Remark 2.6 Hoshino [27] called the multiplicative measure induced by a exponential structure the *limiting conditional compound Poisson distribution*, and characterized it as the limit law of the counts of a sparse contingency table under the law of small numbers.

Example 2.13 (Logarithmic series model) The Ewens sampling formula (2.12) is the multiplicative measure induced by the exponential structure given by $w_x = \theta(x - 1)!$. Here,

$$\mathbb{P}(X_i = x) = \frac{1}{-\log(1 - \xi)} \frac{\xi^x}{x}$$

is the Fisher's logarithmic series distribution [28]. The unconditional distribution of the size index, given by

$$\mathbb{P}(C_1 = c_1, C_2 = c_2 \cdots) = (1 - \xi) \prod_{i \geq 1} \left(\frac{\xi^i}{i}\right)^{c_i} \frac{1}{c_i!}.$$

According to Hoshino and Takemura [29], the Fisher's logarithmic series distribution is the limit distribution of a Poisson-gamma model [30] under the law of small numbers. The discussion parallels derivation of the Ewens sampling formula as the limit of the Dirichlet-multinomial distribution (see Remark 4.6).

Example 2.14 (*Limiting quasi-multinomial distribution*) The limiting quasi-multinomial distribution introduced by Hoshino [31] is the multiplicative measure induced by the exponential structure given by $w_x = \theta x^{x-1}$. Here,

$$\mathbb{P}(X_i = x) = \frac{x^{x-1}}{-W_0(-\xi)} \frac{\xi^x}{x!}.$$

This distribution is the tilted multiplicative measure induced by forests of labeled rooted trees (see Example 2.10). The closed form of the normalization constant, or the Bell polynomial, is an advantage for applications.

Example 2.15 (*Macdonald symmetric functions*) This example is a continuation of Example 2.11. The multiplicative measure induced by the exponential structure associated with the Macdonald symmetric function is the multiplicative measure associated with the exponential structure given by $w_x = (x-1)!(t^x - 1)/(q^x - 1)$. Here,

$$\mathbb{P}(X_i = x) = \left\{ \log \frac{(tx; q)_\infty}{(x; q)_\infty} \right\}^{-1} \frac{t^x - 1}{q^x - 1} \frac{\xi^x}{x}.$$

The mixing of a sampler from this multiplicative measure was discussed by Diaconis and Lam [32], which will be explained in Sect. 5.1.3.

A well-known generalization of the Stirling number is the *generalized factorial coefficient*. Various accounts of the Stirling numbers and their generalizations are given in Chap. 2 of [33]. The generalized factorial is defined as

$$[x]_{i;a} := x(x - a) \cdots (x - (i - 1)a), \qquad i \in \mathbb{N},$$

with $[x]_{0;a} = 1$ by convention. The generalized factorial coefficients $C(n, k; \alpha)$, $\alpha \neq 0$, is defined as identities among the generalized factorials

$$\sum_{k=1}^{n} C\left(n, k; \frac{b}{a}\right) \frac{[t]_{k;b}}{b^k} = \frac{[t]_{n;a}}{a^n}, \qquad a, b \neq 0. \tag{2.23}$$

The *generalized Stirling numbers* [21] are given by

$$S_{n,k}^{a,b} \equiv C\left(n, k; \frac{b}{a}\right) \frac{a^n}{b^k}.$$

For the sequences

$$w_i = (1 - \alpha)_{i-1}, \qquad i \in \mathbb{N}, \qquad v_k = (\theta)_{k;\alpha}, \qquad k \in \mathbb{N} \tag{2.24}$$

with $\alpha < 1$, the negative binomial theorem and Faà di Bruno's formula (2.18) yields

$$B_n((\theta)_{:\alpha}, (1-\alpha)_{.-1}) = \sum_{k=1}^{n} B_{n,k}((1-\alpha)_{.-1})(\theta)_{k;\alpha} = (\theta)_n.$$

Comparing this expression with the definition (2.23), we can relate the partial Bell polynomials to the generalized factorial coefficients as follows:

$$C(n, k; \alpha) = (-1)^n B_{n,k}((-\alpha)_.). \tag{2.25}$$

The following closed form expressions (see p. 158 of [22]) will be useful in later chapters of this monograph:

$$C(n, k; -1) = (-1)^n B_{n,k}(\cdot!) = (-1)^n \frac{n!}{k!} \binom{n-1}{k-1}, \tag{2.26}$$

$$C\left(n, k; \frac{1}{2}\right) = (-1)^{n-k} B_{n,k}\left(\left(-\frac{1}{2}\right)_.\right)$$
$$= (-1)^{n-k} \frac{(n-1)!}{(k-1)!} \binom{2n-k-1}{n-1} 2^{k-2n}, \tag{2.27}$$

where (2.26) is known as the Lah number. Some limits are

$$\lim_{\alpha \to -\infty} \alpha^{-n} C(n, k; \alpha) = S(n, k),$$
$$\lim_{\alpha \to 0} \alpha^{-k} C(n, k; \alpha) = (-1)^{n-k} |s(n, k)|,$$
$$\lim_{\alpha \to 1} C(n, k; \alpha) = (-1)^{n-k} \delta_{n,k}.$$

Example 2.16 (Pitman partitions) The Gibbs partition (2.21) with the sequences (2.24) has the following closed form:

$$\mu_n(\lambda) = \frac{n!(\theta)_{l(\lambda);\alpha}}{(\theta)_n} \prod_{i=1}^{n} \left(\frac{(1-\alpha)_{i-1}}{i!}\right)^{c_i} \frac{1}{c_i!}, \qquad \lambda \in \mathscr{P}_n. \tag{2.28}$$

This expression is known as the Pitman partition [34]. Being a probability mass function, the parameter values of (2.28) are restricted to one of the following sets.

- $\alpha = 0$ and $\theta > 0$ (the Ewens sampling formula (2.12));
- $\alpha < 0$ and $\theta = -m\alpha, m \in \mathbb{N}$ (the m-variate symmetric Dirichlet-multinomial distribution; see Remark 4.5);
- $\alpha \in (0, 1)$ and $\theta > -\alpha$.

For the given length $l(\lambda) = k$ conditional probability measure is given by

$$\mu_{n,k}(\lambda) = \frac{n!(-1)^n}{C(n,k;\alpha)} \prod_{i=1}^{n} \left(\frac{(-\alpha)_i}{i!}\right)^{c_i} \frac{1}{c_i!}, \qquad \lambda \in \mathscr{P}_{n,k}. \qquad (2.29)$$

The Pitman partition will be revisited in Chap. 4.

Remark 2.7 By virtue of the closed form expression of the generalized factorial coefficient of the parameter $\alpha = 1/2$ (2.27), some Gibbs partitions other than the Pitman partition are also closed form. For example, a distribution called the limiting conditional inverse Gaussian-Poisson distribution, derived by Hoshino [35], is a Gibbs partition with the sequences $w_i = (1/2)_{i-1}$ and $v_k = \theta^k$. The normalization constant, or weighted Bell polynomial, takes the form

$$B_n(\theta^\cdot, (1/2)._{-1}) = \frac{2\theta^{n+1/2}e^{2\theta}}{\sqrt{\pi}} K_{n-1/2}(2\theta),$$

where $K_{n-1/2}(\theta)$ is the modified Bessel function of the second kind. Tilted measures involving the generalized factorial coefficient of parameters other than 1/2, are discussed with survey of relevant distribution theory in [36]. Another example is the normalized-inverse Gaussian prior process, which will be introduced in Chap. 4.

2.3   Extremes in Gibbs Partitions

The behavior of extreme sizes of parts in random combinatorial structures is a classical subject. A statistical application is Fisher's exact test for the maximum component in a periodgram [37]. The asymptotic behavior of extreme sizes of parts has also been studied from purely mathematical interest, whose early studies include [38, 39]. In number theory, a number whose largest prime factor does not exceed x is called an *x-smooth number*, whereas a number whose smallest prime factor exceed y is called *y-rough number*. Chapters 3.5 and 3.6 of [40] discuss the smooth and rough numbers, respectively. Interestingly, the limiting distributions of the counting functions of smooth and rough numbers coincide with the limiting distribution of the largest and smallest cycle lengths in the decomposition of permutation into cycles, respectively. This coincidence and other interesting aspects of the relationship between the decomposition of permutation into cycles and integer factorization are presented in Sects. 1.2 and 1.3 of [9] and Sect. 4 of [41].

Random combinatorial structures have been studied by both probabilistic and analytic approaches. A comprehensive exposition of analytic combinatorics is [11]. Panario and Richmond [42] demonstrated that *singularity analysis* of generating functions obtains the asymptotic behavior of the descending ordered cycle lengths in the exp-log class (see Remark 2.2) of exponential structures. The Ewens sampling formula (2.12) is an exp-log class and has several nice properties. For instance, it satisfies the conditioning relation (2.5) and the *logarithmic condition*

$$i\mathbb{P}(X_i = 1) \to \theta, \qquad i\mathbb{E}(X_i) \to \theta, \qquad i \to \infty, \tag{2.30}$$

where $(X_1, X_2, \ldots)$ is the sequence of independent random variables in the conditioning relation. These condition lead to asymptotic independence of small components (Theorem 6.5 of [9]):

$$(C_1, C_2, \ldots) \overset{d}{\to} (X_1, X_2, \ldots), \qquad n \to \infty. \tag{2.31}$$

The behavior was established in the Ewens sampling formula by Arratia et al. [43] and Sibuya [17]. As we will see in Chap. 4, the properties of the limiting distribution of descending ordered proportions of parts, called the *Poisson–Dirichlet distribution*, are well established. The Poisson–Dirichlet distribution is comprehensively surveyed in [44]. The above properties confer great power on probabilistic approaches (see [9] for the extensive discussion). However, probabilistic approaches tend to be model-specific. In this section, we discuss analytic approaches to the asymptotic behaviors of extreme sizes in Gibbs partitions (2.21).

To illustrate singularity analysis of generating functions, let us discuss the signless Stirling number of the first kind, which was discussed in Example 2.9. The ordinary generating function is $(\theta)_n$. By using the Stirling formula

$$\Gamma(z) \sim \sqrt{2\pi z} z^z e^{-z} \{1 + O(z^{-1})\}, \qquad z \to \infty,$$

for $\theta \in \mathbb{C} \backslash \mathbb{Z}_{\leq 0}$, we have

$$\frac{(\theta)_n}{n!} = \frac{\Gamma(\theta + n)}{\Gamma(\theta)\Gamma(n+1)} = \frac{n^{\theta-1}}{\Gamma(\theta)} \{1 + O(n^{-1})\}. \tag{2.32}$$

For a fixed positive integer k, the asymptotic form for $n \to \infty$ is straightforwardly obtained as

$$\frac{|s(n,k)|}{n!} = [\theta^k]\frac{(\theta)_n}{n!} \sim \frac{1}{n}[\theta^{k-1}]\left\{\frac{\exp(\theta \log n)}{\Gamma(\theta + 1)}\right\} = \frac{1}{n}\frac{(\log n)^{k-1}}{(k-1)!}. \tag{2.33}$$

Here, the asymptotic form (2.33) was adequately determined by evaluation of the binomial coefficient (2.32). However, for obtaining a finer result than (2.32), more systematic method is useful. Using complex analysis of the exponential generating function of $(\theta)_n$, $\mathscr{F}(x) = (1-x)^{-\theta}$, Flajolet and Odlyzko [45] established the following result (Proposition 1 of [45]).

Proposition 2.1 ([45]) *For* $\theta \in \mathbb{C} \backslash \mathbb{Z}_{\leq 0}$,

$$[u^n](1-u)^{-\theta} \sim \frac{n^{\theta-1}}{\Gamma(\theta)} \left\{ 1 + \sum_{i \geq 1} \frac{e_i}{n^i} \right\}, \qquad n \to \infty.$$

where $e_i := \sum_{j=i}^{2i} \lambda_{i,j} [\theta - 1]_j$ *with*

$$\sum_{i,j \geq 0} \lambda_{i,j} v^{-i} t^j = e^t \left(1 + \frac{t}{v} \right)^{-(v+1)}.$$

Proof By virtue of the Cauchy-Goursat theorem,

$$[u^n](1-u)^{-\theta} = \frac{1}{2\pi i} \oint \frac{(1-u)^{-\theta}}{u^{n+1}} du, \qquad i := \sqrt{-1}.$$

Take the cut at $[1, \infty)$ and the contour $\mathscr{C} = \gamma_1 \cup \gamma_2 \cup \gamma_3 \cup \gamma_4$, where

$$\gamma_1 = \left\{ u = 1 - \frac{t}{n}; \; t = e^{i\theta}, \; \theta \in \left[\frac{\pi}{2}, -\frac{\pi}{2} \right] \right\},$$

$$\gamma_2 = \left\{ u = 1 + \frac{t+i}{n}; \; t \in [0, n] \right\},$$

$$\gamma_3 = \left\{ u; \; |u| = \sqrt{4 + \frac{1}{n^2}}; \; \Re(u) \leq 2 \right\},$$

$$\gamma_4 = \left\{ u = 1 + \frac{t-i}{n}; \; t \in [0, n] \right\}.$$

The contour is shown in Fig. 2.5. The contribution from γ_3 is exponentially small. With changing variable $u = 1 + t/n$ the contribution from $\mathscr{H} = \gamma_4 \cup \gamma_1 \cup \gamma_2$ is

$$I := \frac{n^{\theta-1}}{2\pi i} \int_{\mathscr{H}} \left(1 + \frac{t}{n} \right)^{-(n+1)} (-t)^{-\theta} dt.$$

Extending the rectilinear parts of the contour $\mathscr{H}$ towards $+\infty$ gives a new contour $\mathscr{H}'$, and the process introduces only exponentially small terms in the integral. The Hankel representation of the gamma function

$$\frac{1}{2\pi i} \oint_{\mathscr{H}'} e^{-x}(-x)^{-z} dx = \frac{1}{\Gamma(z)}$$

yields the leading term. It is the expansion of $(1 + t/n)^{-(n+1)}$ in descending powers of $1/n$ that provides an explicit form of the e_i. $\square$

Fig. 2.5 The contour $\mathscr{C}$
used in the proof of
Proposition 2.1

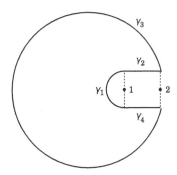

Remark 2.8 Hwang [46] obtained more general result by using singularity analysis of the generating function. He showed that

$$\frac{|s(n,k)|}{n!} \sim \frac{(\log n)^{k-1}}{(k-1)!n}\left[\left\{\Gamma\left(1+\frac{k-1}{\log n}\right)\right\}^{-1} + O\left(\frac{k}{(\log n)^2}\right)\right], \qquad n \to \infty,$$

which holds uniformly over the range $2 \le k \le \eta \log n$ for any $\eta > 0$.

2.3.1 Extremes and Associated Partial Bell Polynomials

To discuss extreme sizes of parts in the exponential structures and the extension, the Gibbs partition (2.21), the *associated partial Bell polynomials* introduced in [47] are useful. They are obtained by restricting the supports to subsets of integer partitions. We call the polynomials *associated* since they cover associated numbers appear in combinatorics literature, such as the associated signless Stirling numbers of the first kind and the associated generalized factorial coefficients [22, 33].

Definition 2.2 ([47]) Suppose the partial Bell polynomial $B_{n,k}(w)$ defined by a sequence of positive real numbers (w_i), $i \in \mathbb{N}$ (2.15). The associated partial Bell polynomials are defined as follows.

$$B_{n,k,(r)}(w) := \sum_{\lambda \vdash n, l(\lambda)=k, c_{<r}(\lambda)=0} n! \prod_{i=1}^{n} \left(\frac{w_i}{i!}\right)^{c_i} \frac{1}{c_i!}, \qquad n \ge rk, \qquad (2.34)$$

$$B_{n,k}^{(r)}(w) := \sum_{\lambda \vdash n, l(\lambda)=k, c_{>r}(\lambda)=0} n! \prod_{i=1}^{n} \left(\frac{w_i}{i!}\right)^{c_i} \frac{1}{c_i!}, \qquad k \le n \le rk, \qquad (2.35)$$

with conventions $B_{n,k,(r)}(w.) = 0$, $n < rk$, $B_{n,k}^{(r)}(w.) = 0$, $n < k$, $n > rk$, and $B_{n,k}^{(r)}(w.) \equiv B_{n,k}(w.)$, $n \le r+k-1$.

Remark 2.9 If we define modified sequences, $w_{(r)i} = w_i, i \geq r, w_{(r)i} = 0, i \in [r - 1]$, and $w_i^{(r)} = w_i, i \in [r], w_i^{(r)} = 0, i > r$, these associated partial Bell polynomials are the partial Bell polynomials $B_{n,k}(w_{(r)})$ and $B_{n,k}(w^{(r)})$, respectively.

Example 2.17 (*associated Striling number of the first kind*) The associated partial Bell polynomial $B_{n,k,(r)}((\cdot - 1)!) \equiv |s_r(n, k)|$ is called the r-associated signless Stirling number of the first kind [33], which is the number of permutations of $[n]$ which consist of k cycles whose lengths are not smaller than r. Other associated Stirling numbers and generalization have similar interpretations.

Example 2.18 (*associated signless Lah number*) The associated partial Bell polynomial $B_{n,k}^{(r)}(\cdot!) \equiv |L_{n,k}^{(r)}|$ is the associated signless Lah number, which appeared in (1.8).

The associated partial Bell polynomials can be expressed in terms of the partial Bell polynomials (Proposition 5 in [47] and Proposition 2.2 in [48]). Let us see an expression for later discussion.

Proposition 2.2 *For positive integer r and k, the associated partial Bell polynomials, $B_{n,k}^{(r)}(w)$, satisfy, if $r + k \leq n \leq rk$,*

$$B_{n,k}^{(r)}(w) = B_{n,k}(w)$$
$$+ \sum_{l=1}^{\lfloor (n-k)/r \rfloor} \frac{(-1)^l}{l!} \sum_{\substack{i_1,\dots,i_l \geq r+1, \\ i_1+\cdots+i_l \leq n-k+l}} B_{n-(i_1+\cdots+i_l),k-l}(w)[n]_{i_1+\cdots+i_l} \prod_{j=1}^{l} \frac{w_{i_j}}{i_j!}$$

and $B_{n,k}^{(r)}(w) = B_{n,k}(w)$ if $k \leq n \leq r + k - 1$.

The associated partial Bell polynomials can be computed by solving recurrence relations, which are given in Appendix A of [47]. One of the recurrence relation is as follows. It also provides the partial Bell polynomials by setting $r \geq n - k + 1$.

Proposition 2.3 *The associated partial Bell polynomials (2.35) satisfy*

$$B_{n+1,k}^{(r)}(w) = \sum_{i=0\vee(n-rk+r)}^{(r-1)\wedge(n-k+1)} \binom{n}{i} w_{i+1} B_{n-i,k-1}^{(r)}(w), \qquad k \leq n+1 \leq rk$$

with convention $B_{j,0}^{(r)}(w) = \delta_{j,0}, \ j \in \mathbb{N}_0$. Here, $a \vee b := \max(a, b)$ and $a \wedge b := \min(a, b)$.

Let us discuss distribution of extreme sizes of parts in the Gibbs partition (2.21). They can be expressed as simple forms in terms of the associated partial Bell polynomials. Let the descending ordered sizes of parts be denoted by $(N_{(1)}, N_{(2)}, \dots, N_{(l(\Lambda))})$. By definitions of the associated partial Bell polynomials and the expression of the conditional probability measure given the length of partition (2.14), we have the conditional distributions of extremes. The largest size follows

$$\mathbb{P}(N_{(1)} \le r|l(\Lambda) = k) = \frac{B_{n,k}^{(r)}(w)}{B_{n,k}(w)}, \qquad \frac{n}{k} \le r \le n, \qquad (2.36)$$

and $\mathbb{P}(N_{(1)} \le r|l(\Lambda) = k) = 0$ for $1 \le r < n/k$, while the smallest size follows

$$\mathbb{P}(N_{(l(\Lambda))} \ge r|l(\Lambda) = k) = \frac{B_{n,k,(r)}(w)}{B_{n,k}(w)}, \qquad 1 \le r \le \frac{n}{k}, \qquad (2.37)$$

and $\mathbb{P}(N_{(l(\Lambda))} \ge r|l(\Lambda) = k) = 0$ for $r > n/k$. Mixing with the distribution of the length of partition (2.22), we obtain the marginal distributions of the extremes. For the largest size, we have

$$\mathbb{P}(N_{(1)} \le r) = \sum_{k=\lceil n/r \rceil}^{n} \frac{v_{n,k} B_{n,k}^{(r)}(w)}{B_n(v, w)} = \frac{B_n(v, w^{(r)})}{B_n(v, w)}, \qquad (2.38)$$

and for the smallest size, we have

$$\mathbb{P}(N_{(l(\Lambda))} \ge r) = \sum_{k=1}^{\lfloor n/r \rfloor} \frac{v_{n,k} B_{n,k,(r)}(w)}{B_n(v, w)} = \frac{B_n(v, w_{(r)})}{B_n(v, w)}, \qquad (2.39)$$

where the sequences $w^{(r)}$ and $w_{(r)}$ are defined in Remark 2.9. Note that the rightmost hand sides of the expressions (2.38) and (2.39) can be computed by use of Faá di Bruno's formula (2.18).

2.3.2 Asymptotics

In this subsection some results on asymptotics of extreme sizes of parts in the Gibbs partition (2.21) will be presented. More results are presented in [47]. We will concentrate on the Gibbs partitions (2.21) with infinite exchangeability (see Sect. 1.1 and Lemma 4.3), because of their special importance in statistical inferences. Nevertheless, the methods used in this subsection will be applicable to Gibbs partitions without infinite exchangeability. As we will see in Chap. 4, the Gibbs partitions with infinite exchangeability can be characterized by the sequence

$$w_i = (1 - \alpha)_{i-1}, \qquad i \in \mathbb{N}, \qquad \alpha < 1,$$

while the sequence $(v_{n,k})$ will be discussed in Chap. 4.

Let us prepare some asymptotic form of the generalized factorial coefficients (2.25). See [47] for the derivations. For fixed positive integer k, we have

$$\frac{C(n, k; \alpha)}{n!} \sim \frac{(-1)^{n-k}}{(k-1)!} \frac{\alpha}{\Gamma(1-\alpha)} n^{-1-\alpha}, \qquad \alpha > 0 \qquad (2.40)$$

and

$$\frac{C(n, k; \alpha)}{n!} \sim \frac{(-1)^n}{\Gamma(-k\alpha)k!} n^{-1-k\alpha}, \qquad \alpha < 0. \qquad (2.41)$$

as $n \to \infty$.

Our discussion in this subsection only needs these results with k fixed. However, for discussion in later chapters, let us prepare some more results on the asymptotics. For $\alpha \in (0, 1)$ an asymptotic form

$$\frac{C(n, k; \alpha)}{n!} \sim \frac{(-1)^{n-k}}{(k-1)!} g_\alpha(s) \alpha n^{-1-\alpha}, \qquad k \sim s n^\alpha, \qquad n \to \infty,$$

follows immediately from the expression for the asymptotics of length of partition (2.46) [49]. Here, $g_\alpha(s)$ is the probability density of the Mittag-Leffler distribution defined in Theorem 2.1. For $\alpha \neq 0$, Keener et al. [50] obtained an asymptotic form for $n \asymp k \to \infty$ using a normal approximation of the distribution of length of partition. Their expression is less explicit, however, useful for numerical computation.

Lemma 2.1 ([50]) *For the generalized factorial coefficient of parameter $\alpha \neq 0$, the generalized factorial coefficient satisfies*

$$\frac{C(n, k; \alpha)}{n!} \sim \frac{1}{k!} \left(\frac{x}{x^* - x} \right)^k \frac{(\alpha x^* - 1)^n}{\sqrt{2\pi n \sigma_*^2}} \sqrt{\frac{\alpha(x^* - x)x}{\alpha x^* - 1}}, \qquad \frac{k}{n} \to x, \qquad (2.42)$$

as $n \asymp r \to \infty$, where

$$\sigma_*^2 := \operatorname{sgn}(\alpha) \frac{x - x^*}{x^{*2}} \left(x - \frac{\alpha(x^* - x)}{\alpha x^* - 1} \right)$$

and x^* solves

$$x = x^* \left\{ 1 - \left(\frac{\alpha x^* - 1}{\alpha x^*} \right)^\alpha \right\}.$$

Here, x^* is the unique positive or negative solution for $\alpha < 0$ and $\alpha > 0$, respectively.

Simple manipulations of the Riemann sums yields asymptotic forms of the associated generalized factorial coefficient

$$C_r(n, k; \alpha) := B_{n,k,(r)}((-\alpha).)(-1)^n$$

and that of the associated signless Stirling number of the first kind

$$|s_r(n, k)| := B_{n,k,(r)}((\cdot - 1)!).$$

For fixed integer k with $2 \le k < n/r$ and non-zero α, we have

$$\frac{C_r(n, k; \alpha)}{n!} \sim \frac{(-1)^n}{\Gamma(-k\alpha)k!} \mathscr{I}_{x,x}^{(k-1)}(-\alpha; -\alpha)n^{-1-k\alpha}, \qquad \frac{r}{n} \to x < \frac{1}{k}, \qquad (2.43)$$

and

$$\frac{|s_r(n, k)|}{n!} \sim \frac{1}{k!} \mathscr{I}_{x,x}^{(k-1)}(0; 0)n^{-1}, \qquad \frac{r}{n} \to x < \frac{1}{k}. \qquad (2.44)$$

as $n \asymp r \to \infty$. Here, we used *incomplete Dirichlet integrals of real parameters* ρ, v, and $m \in \mathbb{N}$

$$\mathscr{I}_{p,q}^{(m)}(v; \rho) := \frac{\Gamma(\rho + mv)}{\Gamma(\rho)\Gamma(v)^m} \int_{\Delta_m(p,q)} y_{m+1}^{\rho-1} \prod_{j=1}^{m} y_j^{v-1} dy_j$$

with convention

$$\mathscr{I}_{p,q}^{(m)}(0; \rho) := \int_{\Delta_m(p,q)} y_{m+1}^{\rho-1} \prod_{j=1}^{m} y_j^{-1} dy_j,$$

where $y_{m+1} := 1 - \sum_{i=1}^{m} y_i$. The support is the interior of the m-dimensional simplex

$$\Delta_m(p, q) := \{(y_1, ..., y_m); p < y_i, i \in [m], q < y_{m+1}\},$$

where

$$0 < q < 1, \qquad 0 < p < \frac{1 - q}{m}.$$

When either of ρ or v is not positive, the integral over all of the simplex does not exist.

The next proposition for the smallest size of parts given the length of partition follows immediately by substituting the asymptotic forms (2.33), (2.40), (2.41), (2.43), and (2.44) into (2.37).

Lemma 2.2 ([47]) *In an infinite exchangeable Gibbs partition, the smallest size of parts given the length of partition satisfies*

$$\mathbb{P}(N_{(l(\Lambda))} \ge r | l(\Lambda) = k) \sim \omega_\alpha(k; x), \qquad \frac{r}{n} \to x < \frac{1}{k}, \qquad 2 \le k < \frac{n}{r},$$

as $n \asymp r \to \infty$ with k fixed, where

$$
\omega_\alpha(k; x) := \begin{cases} \dfrac{1}{k}\mathscr{I}_{x,x}^{(k-1)}(0; 0)(\log n)^{1-k}, & \alpha = 0, \\[2mm] \mathscr{I}_{x,x}^{(k-1)}(-\alpha; -\alpha), & \alpha < 0, \\[2mm] \dfrac{\Gamma(-\alpha)}{\Gamma(-k\alpha)} \dfrac{(-1)^{k-1}}{k} \mathscr{I}_{x,x}^{(k-1)}(-\alpha; -\alpha)n^{-(k-1)\alpha}, & \alpha \in (0,1). \end{cases}
$$

Remark 2.10 The distribution for $\alpha < 0$ is the distribution of the smallest element in the k-variate symmetric Dirichlet distribution. For $\alpha \in [0, 1)$ the smallest size of parts tends to $o(n)$ in probability.

Asymptotics of marginal distributions of extreme sizes of parts in an infinite exchangeable Gibbs partition depend on the choice of the sequence $(v_{n,k})$. Let us take the Pitman partition in Example 2.16 as an example. That is $v_{n,k} = (\theta)_{k;\alpha}$. For the Pitman partition many results on extreme sizes of parts have been obtained by using probabilistic approaches (see Chap. 5 of [9], Sect. 3.3 of [21], and Chaps. 2 and 3 of [44]). Let us see some reproductions of known results and a result obtained in [47] using an analytic approach.

Before proceeding to the results on extreme sizes of parts, let us briefly summarize behavior of the length of the Pitman partition. The probability mass function (2.22) is now

$$
\mathbb{P}(l(\Lambda) = k) = \frac{\theta^k}{(\theta)_n}|s(n,k)|, \qquad \alpha = 0, \qquad \theta > 0,
$$

and

$$
\mathbb{P}(l(\Lambda) = k) = \frac{[-\theta/\alpha]_k}{[-\theta]_n}C(n,k;\alpha), \qquad \alpha \in (0,1), \qquad \theta > -\alpha.
$$

For $\alpha < 0$, $\theta = -m\alpha$, $m \in \mathbb{N}$, the length of partition converges to m almost surely. The asymptotics are given in the next theorem.

Theorem 2.1 ([49, 51]) *In the Pitman partition, for $\alpha = 0$ and $\theta > 0$ the length of partition satisfies*

$$
\lim_{n\to\infty} \frac{l(\Lambda)}{\log n} = \theta, \qquad a.s. \tag{2.45}
$$

For $0 < \alpha < 1$ and $\theta > -\alpha$,

$$
\frac{l(\Lambda)}{n^\alpha} \to S_\alpha, \qquad a.s. \tag{2.46}
$$

as $n \to \infty$. Here, S_α is a strictly positive random variable with continuous density

$$
\mathbb{P}(S_\alpha \in (s, s+ds]) = \frac{\Gamma(1+\theta)}{\Gamma(1+\theta/\alpha)}s^{\frac{\theta}{\alpha}}g_\alpha(s)ds
$$

where $g_\alpha(s)$ is the probability density function of a Mittag-Leffler distribution whose p-th moment is $\Gamma(1+p)/\Gamma(1+p\alpha)$.

Remark 2.11 The density of the Mittag-Leffler distribution satisfies (see Chap. 0 of [21])

$$g_\alpha(s) = \frac{f_\alpha(s^{-1/\alpha})}{\alpha s^{1+1/\alpha}},$$

where $f_\alpha(s)$ is the density of the α-stable distribution [52]

$$f_\alpha(s) = \frac{1}{\pi} \sum_{i \geq 0} \frac{(-1)^{i+1}}{i!} \sin(i\pi\alpha) \frac{\Gamma(1+i\alpha)}{s^{1+i\alpha}} \tag{2.47}$$

The connection between the Pitman partition and the α-stable distribution will be explained in Chap. 4.

Remark 2.12 For $\alpha = 0$ and $\theta > 0$, in addition to the law of large numbers (2.45), Watterson showed the central limit theorem [53]. Namely,

$$\frac{l(\Lambda) - \theta \log n}{\sqrt{\theta \log n}}$$

converges to standard normal as $n \to \infty$. Various generalizations and refinements can be found in [9]. Yamato discussed improvements of the approximation in terms of Edgeworth expansions and Poisson approximations [54, 55]. Hansen [56] established a functional limit theorem describing the counts of cycles up to n^u; the process

$$\frac{\sum_{i=1}^{\lfloor n^u \rfloor} C_i - u\theta \log n}{\sqrt{\theta \log n}}, \qquad u \in [0, 1]$$

converges weakly to a standard Brownian motion in $D[0, 1]$. The $L^2(0, 1)$ convergences of standardized processes for the counts of cycles up to n were also established by Tsukuda [57].

Remark 2.13 Throughout this subsection we assume that θ is fixed. The cases that θ grows with n have also been discussed in [58]. In the case of $\alpha = 0$ and $\theta > 0$, Tsukuda [59] established Poisson approximations of the length of a partition and the counts of small components in this setting. He [60] also showed that if $\theta/n^2 \nrightarrow 0$ as $n \to \infty$ the asymptotic normality of the length of a partition does not hold anymore.

For the largest size of parts, Proposition 2.2 is the key. Some manipulations of the Riemann sums with the asymptotic form (2.41) yields the following result. The proof in [47] seems to be more straightforward than the original proofs: Theorem 2 of [61] and Proposition 20 of [62].

Theorem 2.2 *In the Pitman partition, the largest size of parts satisfies*

$$\mathbb{P}(N_{(1)} \leq r) \sim \rho_{\alpha,\theta}(x) = \sum_{k=0}^{\lfloor x^{-1} \rfloor} \rho_{\alpha,\theta}(k; x), \qquad \frac{r}{n} \to x,$$

as $n \asymp r \to \infty$, where

$$
\rho_{\alpha,\theta}(k;x) := \begin{cases}
\dfrac{(-\theta)^k}{k!} \mathscr{I}_{x,0}^{(k)}(0;\theta), & \alpha = 0,\ \theta > 0, \\[2ex]
(-1)^k \dbinom{m}{k} \mathscr{I}_{x,0}^{(k)}(-\alpha;(k-m)\alpha)\mathbf{1}_{\{mx>1\}}, & \alpha < 0,\ \theta = -m\alpha,\ m \in \mathbb{N}, \\[2ex]
\dfrac{(\theta)_{k;\alpha}}{\alpha^k k!} \mathscr{I}_{x,0}^{(k)}(-\alpha;k\alpha+\theta), & \alpha \in (0,1),\ \theta > -\alpha,
\end{cases}
$$

Remark 2.14 The function $\rho_{0,1}(x^{-1})$ is known as Dickman's function for the frequency of smooth numbers in the number theory [40, 63]. The joint distribution of the descending ordered proportion of parts for $\alpha \in (0,1)$ was obtained by Handa [64] as a property of the two-parameter Poisson–Dirichlet distribution, which will be introduced in Chap. 4.

Remark 2.15 Here we assumed that θ is fixed. However, various interesting results exist for large θ (see Chaps. 7 and 8 of [44]). For example, Griffiths [65] and Handa [64] showed that for $\alpha \in [0,1)$ and $\theta > -\alpha$ the Gumbel distribution appears as the limit law.

Theorem 2.2 shows that the largest size of parts is asymptotically $O(n)$ irrespective of α. In contrast to the behavior of the largest size, the asymptotic behavior of the smallest size heavily depends on α.

Let us begin with seeing the behavior in the regime of $n \asymp r$. By using (2.32), we have an asymptotic form

$$
\frac{v_{n,k}}{B_n(v,w)} = \frac{(\theta)_{k;\alpha}}{(\theta)_n} \sim \frac{(\theta)_{k;\alpha}\Gamma(\theta)}{(n-1)!}n^{-\theta}\{1+O(n^{-1})\} \tag{2.48}
$$

as $n \to \infty$ for fixed θ. Substituting the asymptotic forms (2.43), (2.44), and (2.48) into (2.39), we have the following result.

Theorem 2.3 *In the Pitman partition, for $\alpha = 0$ and $\theta > 0$ the smallest size of parts satisfies*

$$
\mathbb{P}(N_{(l(\Lambda))} \geq r) \sim \Gamma(\theta)(xn)^{-\theta}\omega_\theta(x), \qquad \frac{r}{n} \to x,
$$

as $n \asymp r \to \infty$, where

$$
\omega_\theta(x) := x^\theta \sum_{k=1}^{\lfloor x^{-1}\rfloor} \frac{\theta^k}{k!} \mathscr{I}_{x,x}^{(k-1)}(0;0).
$$

For $\alpha < 0$ and $\theta = -m\alpha$, $m \in \mathbb{Z}$,

$$
\mathbb{P}(N_{(l(\Lambda))} \geq r) \sim \mathscr{I}_{x,x}^{(m-1)}(-\alpha;-\alpha)\mathbf{1}_{\{mx<1\}}, \qquad \frac{r}{n} \to x,
$$

as $n \asymp r \to \infty$. For $\alpha \in (0, 1)$ and $\theta > -\alpha$,

$$\mathbb{P}(N_{(l(\Lambda))} \geq r) \sim \frac{\Gamma(1 + \theta)}{\Gamma(1 - \alpha)} n^{-\theta - \alpha}, \qquad n \asymp r \to \infty.$$

Remark 2.16 This result for the case of $\alpha = 0$ and $\theta > 0$ was obtained by Panario and Richmond in Theorem 3 of [42] by an analytic combinatorics approach different from the method here. Moreover, they obtained marginal distributions of the descending ordered sizes of parts. Arratia et al. (Lemma 5.5 of [9]) obtained this result by a probabilistic approach using the conditioning relation (2.5) and the logarithmic condition (2.30).

Remark 2.17 The function $\omega_1(x^{-1})$ is known as Buchstab's function for the frequency of rough numbers in number theory [40, 66].

According to Theorem 2.3, the smallest size of parts is $o(n)$ in probability if $\alpha \in [0, 1)$. Moreover, the distribution degenerates for $\alpha \in (0, 1)$ and $\theta > -\alpha$. In contrast to the case of the largest size of parts, we have no analog of Buchstab's function for $\alpha \in (0, 1)$ and $\theta > -\alpha$, namely, the partition is always "smooth". More precise estimate is inevitable for further discussion of the smallest size of part. For $\alpha = 0$ and $\theta > 0$, the asymptotic independence (2.31) immediately gives the next theorem [67]. This result was obtained by Panario and Richmond in Theorem 3 of [42].

Theorem 2.4 *In the Ewens sampling formula, the smallest size of parts satisfies*

$$\mathbb{P}(N_{(l(\Lambda))} \geq r) \sim e^{-\theta h_{r-1}}, \qquad n \to \infty, \qquad r = o(n) \geq 2,$$

where $h_r = \sum_{i=1}^{r} i^{-1}$. Moreover, $\mathbb{P}(l(\Lambda) \geq r) \sim r^{-\theta} e^{-\gamma \theta}$ as $r \to \infty$, where γ is the Euler-Mascheroni constant.

For $\alpha \in (0, 1)$ and $\theta > -\alpha$ the Pitman partition is not in the exp-log class and does not satisfy the conditioning relation (2.5). Nevertheless, the following theorem can be proven using singularity analysis, as the proof for Proposition 2.1.

Theorem 2.5 ([47]) *In the Pitman partition with $\alpha \in (0, 1)$ and $\theta > -\alpha$, the smallest size of parts satisfies*

$$\mathbb{P}(N_{(l(\Lambda))} \geq r) \sim \frac{\Gamma(1 + \theta)}{\Gamma(1 - \alpha)} \left\{ \sum_{i=1}^{r-1} p_\alpha(i) \right\}^{-1 - \frac{\theta}{\alpha}} n^{-\theta - \alpha}, \qquad n \to \infty,$$

for $r = o(n) \geq 2$, where

$$p_\alpha(i) := \binom{\alpha}{i} (-1)^{i+1}, \qquad i \in \mathbb{N} \tag{2.49}$$

is probability mass function of Sibuya's distribution [68, 69]. Moreover,

$$\mathbb{P}(N_{(l(\Lambda))} \geq r) \sim \frac{\Gamma(1+\theta)}{\Gamma(1-\alpha)} n^{-\theta-\alpha}, \qquad r \to \infty.$$

Remark 2.18 Theorem 2.5 implies that the smallest size of parts converges to one in probability. This theorem and Theorem 2.3 tell us more. Since

$$\mathbb{P}(N_{(l(\Lambda))} = n) \sim \frac{\Gamma(1+\theta)}{\Gamma(1-\alpha)} n^{-\theta-\alpha}, \qquad n \to \infty,$$

we have $\mathbb{P}(r \leq l(\Lambda) < n) = o(n^{-\theta-\alpha})$ as $n, r \to \infty, r \asymp n$. Therefore, apart from the mass at n, the probability mass concentrates around $o(n)$.

Remark 2.19 It is known that asymptotic behavior of smallest sizes of part in an infinite exchangeable random partition has universal properties. The early works are [70, 71]. Lemma 3.11 of [21] is as follows. The scaled length of partition

$$T_\alpha := \lim_{n\to\infty} l(\Lambda) n^{-\alpha}, \qquad \alpha \in (0, 1).$$

has the almost sure limit, which is strictly positive and finite, if and only if the decreasing ordered proportion of parts $(P_{(1)}, P_{(2)}, ...)$ satisfies

$$P_{(i)} \sim Z i^{-1/\alpha}, \qquad i \to \infty,$$

for a positive random variable $0 < Z < \infty$, where $Z^{-\alpha} = \Gamma(1-\alpha)T_\alpha$ and the joint distribution of size indices satisfy

$$(C_1, C_2, ...) \sim (p_\alpha(1), p_\alpha(2), ...)T_\alpha n^\alpha, \qquad n \to \infty, \qquad (2.50)$$

where $p_\alpha(j)$ is the Sibuya's distribution (2.49). Theorem 2.1 implies that for the Pitman partition we identify the limiting random variable as $T_\alpha = S_\alpha$. Yamato and Sibuya [72] obtained the joint distribution (2.50) by computing the moments.

References

1. Logan, B.F., Shepp, L.A.: A variational problem for random Young tableaux. Adv. Math. **26**, 206–222 (1977)
2. Vershik, A.M., Kerov, S.V.: Asymptotic behavior of the Plancherel measure of the symmetric group and the limit from of Young tableaux. Dokl. Akad. Nauk SSSR **233**, 1024–1037 (1977)
3. Kerov, S.V.: Asymptotic Representation Theory of the Symmetric Group and its Applications in Analysis, vol. 219. American Mathematical Society, Providence (2003). Translations of Mathematical Monograph
4. Borodin, A., Corwin, I.: Macdonald processes. Probab. Theor. Relat. Fields **158**, 225–400 (2014)

5. Vershik, A.M.: Statistical mechanics of combinatorial partitions, and their limit configurations. Funct. Anal. Appl. **30**, 90–105 (1996)
6. Green, M.B., Schwarz, J.H., Witten, E.: Superstring Theory, vol. 2. Cambridge University Press, New York (1987)
7. Nekrasov, N.A., Okounkov, A.: Seiberg-Witten theory and random partitions. In: Etingof, P., Retakh, V., Singer, I.M. (eds.) The Unity of Mathematics. Progress in Mathematics, vol. 244, pp. 525–596. Birkhäuser, Boston (2006)
8. Takemura, A., et al.: Special issue: statistical theory of statistical disclosure control problem. Proc. Inst. Stat. Math. **51**, 181–388 (2003)
9. Arratia, R., Barbour, A.D., Tavaré, S.: Logarithmic Combinatorial Structures: A Probabilistic Approach. European Mathematical Society, Zürich (2003). EMS Monographs in Math
10. Stanley, R.P.: Enumerative Combinatorics, vol. 2. Cambridge University Press, New York (1999)
11. Flajolet, P., Sedgewick, R.: Analytic Combinatorics. Cambridge University Press, Cambridge (2009)
12. Flajolet, P., Soria, M.: Gaussian limiting distributions for the number of components in combinatorial structures. J. Combin. Theor. Ser. A **53**, 165–182 (1990)
13. Corless, R., Gonnet, G., Hare, D., Jeffrey, D., Knuth, D.: On the Lambert W function. In: Advances in Computational Mathematics, vol. 5, pp. 329–359. Springer, Heidelberg (1996)
14. Katz, L.: Probability of indecomposability of a random mapping function. Ann. Math. Stat. **26**, 512–517 (1955)
15. Ewens, W.J.: The sampling theory of selectively neutral alleles. Theor. Popul. Biol. **3**, 87–112 (1972)
16. Antoniak, C.: Mixture of Dirichlet processes with applications to Bayesian nonparametric problems. Ann. Statist. **2**, 1152–1174 (1974)
17. Sibuya, M.: A random-clustering process. Ann. Inst. Stat. Math. **45**, 459–465 (1993)
18. Tavaré, S., Ewens, W.J.: The Ewens sampling formula. In: Johnson, N.L., Kotz, S., Balakrishnan, N. (eds.) Multivariate Discrete Distributions. Wiley, New York (1997)
19. Crane, H.: The ubiquitous Ewens sampling formula. Stat. Sci. **31**, 1–19 (2016)
20. Bell, E.T.: Exponential polynomials. Ann. Math. **35**, 258–277 (1934)
21. Pitman, J.: Combinatorial stochastic processes. Ecole d'Été de Probabilités de Saint Flour, Lecture Notes in Mathematics, vol. 1875. Springer, Heidelberg (2006)
22. Comtet, L.: Advanced Combinatorics. Ridel, Dordrecht (1974)
23. Macdonald, I.G.: Symmetric Functions and Hall Polynomials, 2nd edn. Oxford University Press, New York (1995)
24. Okano, T., Okuto, Y., Shimizu, A., Niikura, Y., Hashimoto, Y., Yamada, H. The formula of Faà di Bruno and its applications I. Annual Review 2000, Institute of National Sciences, Nagoya City University, pp. 35–44 (2000). in Japanese
25. Kolchin, V.F.: A problem on the distribution of particles among cells, and cycles of random permutations. Teor. Veroyatnost. i Primenen. **16**, 67–82 (1971)
26. Kerov, S.V: Coherent random allocations, and the Ewens-Pitman formula. Zap. Nauchn. Semi. POMI **325**, 127–145 (1995); English translation: J. Math. Sci. **138**, 5699–5710 (2006)
27. Hoshino, N.: Random partitioning over a sparse contingency table. Ann. Inst. Stat. Math. **64**, 457–474 (2012)
28. Fisher, R.A., Corbet, A.S., Williams, C.B.: The relationship between the number of species and the number of individuals in a random sample of an animal population. J. Animal Ecol. **12**, 42–58 (1943)
29. Hoshino, N., Takemura, A.: Relationship between logarithmic series model and other super-population models useful for microdata discrosure risk assesment. J. Jpn. Stat. Soc. **28**, 125–134 (1998)
30. Bethlehem, J.G., Keller, W.J., Pannekoek, J.: Disclosure control of microdata. J. Am. Stat. Assoc. **85**, 38–45 (1990)
31. Hoshino, N.: On a limiting quasi-multinomial distribution. Discussion Paper CIRJE-F-361, Center for International Research on the Japanese Economy, Faculty of Economics, The University of Tokyo (2005)

32. Diaconis, P., Lam, A.: A probabilistic interpretation of the Macdonald polynomials. Ann. Probab. **40**, 1861–1896 (2012)
33. Charalambides, C.A.: Combinatorial Methods in Discrete Distributions. Wiley, New Jersey (2005)
34. Pitman, J.: Exchangeable and partially exchangeable random partitions. Probab. Theor. Relat. Fields **102**, 145–158 (1995)
35. Hoshino, N.: A discrete multivariate distribution resulting from the law of small numbers. J. Appl. Probab. **43**, 852–866 (2006)
36. Hoshino, N.: Engen's extended negative binomial model revisited. Ann. Inst. Stat. Math. **57**, 369–387 (2005)
37. Fisher, R.A.: Tests of significance in harmonic analysis. Proc. Roy. Soc. Lond. Ser. A **125**, 54–59 (1929)
38. Goncharov, V.L.: Some facts from combinatorics. Izv. Akad. Nauk SSSR, Ser. Mat. **8**, 3–48 (1944)
39. Shepp, L.A., Lloyd, S.P.: Ordered cycle length in random permutation. Trans. Am. Math. Soc. **121**, 340–357 (1966)
40. Tenenbaum, G.: Introduction to Analytic and Probabilistic Number Theory. Cambridge University Press, New York (1955)
41. Billingsley, P.: Convergence of Probability Measures, 2nd edn. Wiley, New Jersey (1999)
42. Panario, D., Richmond, B.: Smallest components in decomposable structures: exp-log class. Algorithmica **29**, 205–226 (2010)
43. Arratia, R., Barbour, A.D., Tavaré, S.: Poisson process approximations for the Ewens sampling formula. Ann. Appl. Probab. **2**, 519–535 (1992)
44. Feng, S.: The Poisson–Dirichlet Distributions and Related Topics. Springer, Heidelberg (2010)
45. Flajolet, P., Odlyzko, A.: Singularity analysis of generating functions. SIAM J. Discrete Math. **3**, 216–240 (1990)
46. Hwang, H.-K.: Asymptotic expansions for Stirling's number of the first kind. J. Combin. Theor. Ser. A **71**, 343–351 (1995)
47. Mano, S.: Extreme sizes in the Gibbs-type random partitions. Ann. Inst. Stat. Math. **69**, 1–37 (2017)
48. Mano, S.: Partition structure and the A-hypergeometric distribution associated with the rational normal curve. Electron. J. Stat. **11**, 4452–4487 (2017)
49. Pitman, J.: Brownian motion, bridge, excursion and meander characterized by sampling at independent uniform times. Electron. J. Probab. **4**, 33pp (1999)
50. Keener, R., Rothman, E., Starr, N.: Distribution of partitions. Ann. Stat. **15**, 1466–1481 (1978)
51. Korwar, R.M., Hollander, M.: Contribution to the theory of Dirichlet process. Ann. Probab. **1**, 705–711 (1973)
52. Pollard, H.: The representation of e^{-x^λ} as a Laplace integral. Bull. Am. Math. Soc. **52**, 908–910 (1946)
53. Watterson, G.A.: The sampling theory of selectively neutral alleles. Adv. Appl. Probab. **6**, 463–488 (1974)
54. Yamato, H.: Edgeworth expansions for the number of distinct components associated with the Ewens sampling formula. J. Jpn. Stat. Soc. **43**, 17–28 (2013)
55. Yamato, H.: Poisson approximations for sum of Bernoulli random variables and its application to Ewens sampling formula. J. Jpn. Stat. Soc. **47**, 187–195 (2018)
56. Hansen, J.C.: A functional central limit theorem for the Ewens sampling formula. J. Appl. Probab. **27**, 28–43 (1990)
57. Tsukuda, K.: Functional central limit theorems in $L^2(0, 1)$ for logarithmic combinatorial assemblies. Bernoulli **24**, 1033–1052 (2018)
58. Feng, S.: Large deviations associated with Poisson–Dirichlet distribution and Ewens sampling formula. Ann. Appl. Probab. **17**, 1570–1595 (2007)
59. Tsukuda, K.: Estimating the large mutation parameter of the Ewens sampling formula. J. Appl. Probab. **54**, 42–54 (2017)

60. Tsukuda, K.: On Poisson approximations for the Ewens sampling formula when the mutation parameter grows with the sample size. arXiv: 1704.06768
61. Griffiths, R.C.: On the distribution of points in a Poisson process. J. Appl. Probab. **25**, 336–345 (1988)
62. Pitman, J., Yor, M.: The two-parameter Poisson–Dirichlet distribution derived from a stable subordinator. Ann. Probab. **25**, 855–899 (1997)
63. Dickman, K.: On the frequency of numbers containing prime factors of a certain relative magnitude. Ark. Mat., Astronomi och Fysik. **22**, 1–14 (1930)
64. Handa, K.: The two-parameter Poisson–Dirichlet point process. Bernoulli **15**, 1082–1116 (2009)
65. Griffiths, R.C.: On the distribution of allele frequencies in a diffusion model. Theor. Popul. Biol. **15**, 140–158 (1979)
66. Buchstab, A.A.: An asymptotic estimation of a general number-theoretic function. Mat. Sb. **44**, 1239–1246 (1937)
67. Arratia, R., Tavaré, S.: Limit theorems for combinatorial structures via discrete process approximations. Random Structures Algorithms **3**, 321–345 (1992)
68. Sibuya, M.: Generalized hypergeometric, digamma and trigamma distributions. Ann. Inst. Stat. Math. **31**, 373–390 (1979)
69. Devroye, L.: A triptych of discrete distributions related to the stable law. Stat. Probab. Lett. **18**, 349–351 (1993)
70. Karlin, S.: Central limit theorems for certain infinite urn schemes. J. Math. Mech. **17**, 373–401 (1967)
71. Rouault, A.: Lois de Zipf et sources markoviennes. Ann. Inst. H. Poincaré Sect. B **14**, 169–188 (1978)
72. Yamato, H., Sibuya, M.: Moments of some statistics of Pitman sampling formula. Bull. Inf. Cybernet. **32**, 1–10 (2000)

Chapter 3
A-Hypergeometric Systems

Abstract This chapter introduces the *A*-hypergeometric system of linear partial differential equations. It presents the known results of the *A*-hypergeometric system with a two-row matrix *A* and explains its relationship with integer partitions. The *A*-hypergeometric distribution is a class of discrete exponential families whose normalization constant is the *A*-hypergeometric polynomial. The *A*-hypergeometric distribution emerges in multinomial sampling of log-affine models and is conditional on sufficient statistics. After presenting the properties of the *A*-hypergeometric distribution, the chapter discusses the maximum likelihood estimation of the *A*-hypergeometric distribution of the two-row matrix *A*. Especially, using the properties of partition polytopes, it proves the nonexistence theorem of the maximum likelihood estimator. Finally, it introduces holonomic gradient methods (HGMs), which numerically solve holonomic systems without combinatorial enumeration, and applies a difference HGM to the *A*-hypergeometric polynomials of a two-row matrix *A*.

Keywords *A*-hypergeometric system · Discrete exponential family
Gröbner basis · Holonomic gradient method · Holonomic system
Integer partition · Maximum likelihood estimation · Monomial curve
Newton polytope · Partition polytope

3.1 Preliminaries

This section prepares materials concerning *A*-hypergeometric systems to be used in later discussion. A lucid and comprehensive discussion of *A*-hypergeometric systems is provided in [1]. Backgrounds of computational algebra are given in [2–4], and various statistical applications can be found in [4–7].

Consider a nonnegative integer-valued $d \times m$ matrix A. If there exists a vector c for which $c^\top A = (1, ..., 1)$, we say A is *homogeneous*. In other words, the row span of A contains the vector $(1, ..., 1)$. Throughout this monograph, we will assume that A is homogeneous. The polynomial ring in x over $\mathbb{C}$ is denoted by $\mathbb{C}[x]$. The *toric ideal* is defined as follows.

© The Author(s) 2018   45
S. Mano, *Partitions, Hypergeometric Systems, and Dirichlet Processes in Statistics*,
JSS Research Series in Statistics, https://doi.org/10.1007/978-4-431-55888-0_3

Definition 3.1 The binomial ideal

$$I_A = \left\langle x^{z^+} - x^{z^-} ; z \in \mathrm{Ker} A \cap \mathbb{Z}^m \right\rangle \subset \mathbb{C}[x],$$

where $x^z = \prod_i x_i^{z_i}$, $z_i^+ := \max(z_i, 0)$, and $z_i^- := \max(-z_i, 0)$, is called the *toric ideal* of A.

The zero set of I_A determines a toric variety of dimension $d - 1$ in projective space $\mathbb{P}^{m-1}$.

The reduced Gröbner basis of a toric ideal can be computed using the following algorithm (Algorithm 4.5 of [2]) based on the Buchberger's algorithm.

Algorithm 3.1 ([2]) Let the toric ideal J_A of a polynomial ring $\mathbb{C}[x, t]$ of variables $x_1, x_2, ..., x_n$ and $t_1, t_2, ..., t_d$ be

$$J_A = \langle x_1 - t^{a_1}, x_2 - t^{a_2}, ..., x_n - t^{a_n} \rangle, \qquad t^{a_i} := \prod_{j=1}^{d} t_j^{(a_i)_j},$$

where a_i is the i-th column vector of an integer-valued matrix A. Let G be the reduced Gröbner basis of J_A with respect to an elimination term order $\prec$ with $\{t.\} \succ \{x.\}$. Then, the set $G \cap \mathbb{C}[x]$ is the reduced Gröbner basis of the toric ideal I_A of the polynomial ring $\mathbb{C}[x]$ with respect to $\prec$.

Example 3.1 Let us obtain a reduced Gröbner basis of the matrix

$$A = \begin{pmatrix} 0 & 1 & 2 & 3 \\ 1 & 1 & 1 & 1 \end{pmatrix}.$$

Consider the toric ideal J_A of a polynomial ring $\mathbb{C}[x, t]$ of variables x_1, x_2, x_3, x_4, t_1, t_2;

$$J_A = \langle f_1 = x_1 - \underline{t_2}, f_2 = x_2 - \underline{t_1 t_2}, f_3 = x_3 - \underline{t_1^2 t_2}, f_4 = x_4 - \underline{t_1^3 t_2} \rangle.$$

We use the elimination term order $\prec$ by the reverse lexicographic term order with $t_1 \succ t_2 \succ x_1 \succ x_2 \succ x_3 \succ x_4$. The underlined terms in the above expression are the initial monomials. The Gröbner basis with respect to $\prec$ can be computed by Buchberger's algorithm. We obtain the reduced Gröbner basis of J_A, and the reduced Gröbner basis of the toric ideal I_A is $\{x_1 x_4 - x_2 x_3, x_2 x_4 - x_3^2, x_1 x_3 - x_2^2\}$.

The *Weyl algebra* of dimension m is the free associative $\mathbb{C}$-algebra

$$D := \mathbb{C}\langle x_1, ..., x_m, \partial_1, ..., \partial_m \rangle$$

modulo the commutation rules

$$x_i x_j = x_j x_i, \ \partial_i \partial_j = \partial_j \partial_i, \ \partial_i x_j = x_j \partial_i \text{ for } i \neq j, \text{ and } \partial_i x_i = x_i \partial_i + 1.$$

Gel'fand et al. [8] defined a class of functions called *GKZ-hypergeometric functions*, also referred to as *A-hypergeometric functions*.

Definition 3.2 Let A be a nonnegative integer-valued $d \times m$ matrix of rank d, and let $b \in \mathbb{C}^d$ be a fixed vector. The *A-hypergeometric system* $H_A(b)$ is the following system of linear partial differential equations for an indeterminate function $f(x)$:

$$L_i := \sum_{j=1}^{m} a_{ij}\theta_j - b_i, \qquad i \in [d], \tag{3.1}$$

$$\partial^{c^+} - \partial^{c^-}, \qquad c^+ - c^- \in \ker A \cap \mathbb{Z}^m, \tag{3.2}$$

where $\theta_j := x_j \partial_j$ (the Euler derivative). We regard $H_A(b)$ as a left ideal in the Weyl algebra D and call it the *A-hypergeometric ideal*. The second group of annihilators generates the toric ideal of A.

Definition 3.3 The series solution of the A-hypergeometric function around the origin, given by

$$Z_A(b; x) := \sum_{\{c; Ac = b, c \in \mathbb{N}^m\}} \frac{x^c}{c!}, \qquad x^c := \prod_{i=1}^{m} x_i^{c_i}, \qquad c! := \prod_{i=1}^{m} c_i! \tag{3.3}$$

is called the *A-hypergeometric polynomial*. As a convention, we set $Z_A(b; x) = 0$ if $b \notin A\mathbb{N}_0^m =: \mathbb{N}_0 A$.

The A-hypergeometric system is a *holonomic system* and the A-hypergeometric ideal $H_A(b)$ is a *holonomic ideal*. The definitions and backgrounds around holonomic systems are presented in Sect. 1.4 of [1]. In the following discussion, it can be more efficient to replace the Weyl algebra by the ring of differential operators

$$R := \mathbb{C}(x_1, \ldots, x_m)\langle \partial_1, \ldots, \partial_m \rangle.$$

This is the $\mathbb{C}$-algebra subject to the commutative relations

$$\partial_i \bullet c(x) = c(x) \bullet \partial_i + \frac{\partial c(x)}{\partial x_i}, \qquad c(x) \in \mathbb{C}(x).$$

For a holonomic ideal I, RI is a zero-dimensional ideal in R and $\mathrm{rank}(I) = \dim_{\mathbb{C}}(R/RI)$ (Corollary 1.4.14 in [1]). A Gröbner basis of RI with respect to any term order on R determines a set of *standard monomials*,

$$\{\partial\text{-monomials not in } \mathrm{in}_{\prec}(RI)\} = \{\partial_{p(1)}, \ldots, \partial_{p(\mathrm{rank}(I))}\}, \qquad p \in \mathbb{N}_0^m.$$

Here, we fix the term order as reverse lexicographic with $\partial_1 \succ \partial_2 \succ \cdots \succ \partial_m$. In this monograph, the next result (Theorem 1.4.19 of [1]) is needed.

Theorem 3.1 *Let I be a holonomic ideal and U is a simply connected domain in $\mathbb{C}\setminus\text{Sing}(I)$, where $\text{Sing}(I)$ is the singular locus of I. Consider the system of differential equations $I \bullet f = 0$, i.e., $l \bullet f = 0$, $l \in I$, for holomorphic functions f on U. The dimension of the complex vector space of solutions is equal to $\text{rank}(I)$.*

According to this theorem, the number of independent solutions of an A-hypergeometric system is $\text{rank}(H_A(b))$. The totality of the standard monomials provides a solution basis of the A-hypergeometric system. The lower bound of the rank is known (Theorem 3.5.1 of [1]).

Theorem 3.2 ([1]) *Let A be a homogeneous integer-valued $d \times m$ matrix of rank $d \geq 2$. For any vector $b \in \mathbb{C}^d$, $\text{rank}(H_A(b)) \geq \text{vol}(A)$.*

Other independent solutions can be systematically obtained by perturbing the vector b. This method generalizes the method of Frobenius for finding series solutions of second-order ordinary differential equations (Chap. VI, supplementary note I of [9]). The procedure, which is described in Sects. 3.4 and 3.5 of [1], in Sect. 6.12 of [4], and used in the proof of Theorem 3.2 in [1], is briefly summarized here. The A-hypergeometric system is formally solved as

$$\phi_v(x) = \sum_{\{u;Au=0,u\in\mathbb{Z}^m\}} \frac{[v]_{u_-}}{[u+v]_{u_+}} x^{v+u},$$

where $[v]_{u_-} := \prod_{i;u_i<0}[v_i]_{-u_i}$ and $[u+v]_{u_+} := \prod_{i;u_i>0}[v_i+u_i]_{u_i}$. Here, $v \in (\mathbb{C}\setminus\mathbb{Z}_{<0})^m$ are fake exponents of $H_A(b)$ with respect to a weight vector $w \in \mathbb{R}^n$. These exponents satisfy $Av = b$ and $\prod_i \partial_i^{e_i} x^v = 0, \forall \prod_i \partial_i^{e_i} \in \text{in}_w(I_A)$, where $\text{in}_w(I_A)$ is the initial ideal of the toric ideal I_A with respect to w (see Proposition 3.1.5 in [1] for the definition). In particular, when $b \in \mathbb{N}_0 A$, there is a unique exponent v in $\mathbb{N}_0^m$ and $\phi_v(x)$ is a constant multiple of the A-hypergeometric polynomial (Lemmas 3.4.9 and 3.4.10 in [1]). Choosing a generic vector b' in $\mathbb{C}^d$, we then write $v + \varepsilon v'$ as the corresponding exponent of $H_A(b + \varepsilon b')$. The hypergeometric series

$$\phi_{v+\varepsilon v'}(x) = \sum_{\{u:Au=0,u\in\mathbb{Z}^m\}} \frac{[v+\varepsilon v']_{u_-}}{[u+v+\varepsilon v']_{u_+}} x^{u+v+\varepsilon v'}$$

provides an independent solution which is annihilated by $H_A(b)$.

3.1.1 Results of a Two-Row Matrix

A homogeneous two-row matrix A admits many explicit results for an A-hypergeometric system. This subsection summarizes the results used in the next section, following the presentation in [10]. Many of the results were obtained by Cattani et al. [11] and extensively discussed in Sect. 4.2 of [1].

Let $0 < i_1 < i_2 < \cdots < i_{m-1}$ be relatively prime integers (the greatest common divisor is one). Without loss of generality, we can assume the following form of the two-row matrix A:

$$A = \begin{pmatrix} 0 & i_1 & i_2 & \cdots & i_{m-1} \\ 1 & 1 & 1 & \cdots & 1 \end{pmatrix}, \qquad m \geq 3. \tag{3.4}$$

The convex hull of the column vectors is a one-dimensional polytope with a volume vol(A) of i_{m-1}. The toric ideal I_A determines a degree i_{m-1} *monomial curve* (*toric curve*) in the projective space $\mathbb{P}^{m-1}$. The monomial curve is called normal if and only if $i_{m-1} = m - 1$. In this case, the monomial curve is the embedding of $\mathbb{P}^1$ in $\mathbb{P}^{m-1}$ and is called the *rational normal curve*. The rational normal curve is the image of a Veronese embedding $\mathbb{P}^1 \to \mathbb{P}^{m-1}$ which acts on the homogeneous coordinates $(s : t)$ as

$$(s : t) \mapsto (s^{m-1} : s^{m-2}t : \ldots : t^{m-1}).$$

The algebraic geometry is explained in [12].

Example 3.2 The toric ideal I_A in Example 3.1 determines the rational normal curve, which is specifically called the twisted cubic. This curve is the zero set of the three homogeneous polynomials $x_1 x_4 - x_2 x_3$, $x_2 x_4 - x_3^2$, and $x_1 x_3 - x_2^2$.

Note that a homogeneous two-row matrix generates integer partitions. The matrix A in (3.4) with $i_{m-1} = m - 1$ and the vector $b \in \mathbb{N}_0 A$ determine an integer partition because $Ac = b$ is equivalent to

$$c_1 + \cdots + c_m = b_2, \qquad 1 \cdot c_1 + 2 \cdot c_2 + \cdots + m \cdot c_m = b_1 + b_2,$$

which are the requisite conditions of the size index $c_i := \#\{j; \lambda_j = i\}$ of the partition $\lambda \vdash (b_1 + b_2)$ with $l(\lambda) = b_2$. In general, (3.4) determines the partitions with forbidden summands (see Sect. 2.6 of [13]) satisfying

$$1 \cdot c_1 + (i_1 + 1)c_{i_1+1} + \cdots + (i_{m-1} + 1)c_{i_m+1} = b_1 + b_2$$

with $l(\lambda) = b_2$. The associated partial Bell polynomials $B_{n,k}^{(r)}(w)$ defined in Definition 2.2 are equal to $n!$ times the A-hypergeometric polynomial with

$$A = \begin{pmatrix} 0 & 1 & 2 & \cdots & (r-1) \wedge (n-k) \\ 1 & 1 & 1 & \cdots & 1 \end{pmatrix}, \qquad b = \begin{pmatrix} n-k \\ k \end{pmatrix}, \tag{3.5}$$

where $x_i = w_i/i!$. However, the identity of the associated partial Bell polynomials $B_{n,k,(r)}(w)$ is not evident. Nevertheless, it can be shown [10] that they are proportional to the A-hypergeometric polynomial with

$$A = \begin{pmatrix} 0 & 1 & 2 & \cdots & n - kr \\ 1 & 1 & 1 & \cdots & 1 \end{pmatrix}, \qquad b = \begin{pmatrix} n - kr \\ k \end{pmatrix}.$$

The rank of the A-hypergeometric ideal of a two-row matrix is completely determined. The following theorem is Theorem 3.7 in [11] and Theorem 4.2.4 in [1].

Theorem 3.3 ([1, 11]) *For a homogeneous two-row matrix A, the rank of the A-hypergeometric ideal is*

$$\operatorname{rank}(H_A(b)) = \begin{cases} \operatorname{vol}(A), & b \notin \mathscr{E}(A), \\ \operatorname{vol}(A) + 1, & b \in \mathscr{E}(A), \end{cases}$$

where

$$\mathscr{E}(A) := ((\mathbb{N}_0 A + \mathbb{Z}a_1) \cap (\mathbb{N}_0 A + \mathbb{Z}a_m)) \backslash \mathbb{N}_0 A, \qquad \mathbb{Z}a_i := \{\lambda a_i : \lambda \in \mathbb{Z}\}.$$

An explicit solution basis for an arbitrary A-hypergeometric system with a two-row matrix is given in [11]. The following lemma (Proposition 1.2 and Lemma 1.3 in [11]) is useful for our discussion.

Lemma 3.1 ([11]) *For a homogeneous two-row matrix A, let φ be a local holomorphic solution of the A-hypergeometric system $H_A(b)$. If φ is a polynomial, then it is a Laurent polynomial. Moreover, when the vector b is $b \in \mathbb{N}_0 A$, the only Laurent polynomials are the constant multiples of the A-hypergeometric polynomial (3.3).*

Example 3.3 For a matrix A (3.4) with $i_{m-1} = m - 1$, $\mathscr{E}(A) = \phi$ and $\operatorname{rank}(H_A(b)) = \operatorname{vol}(A) = m - 1$.

Example 3.4 Let us consider an example with $\mathscr{E}(A) \neq \phi$. The same example was presented as Example 4.3.9 in [1], one of Examples 1.8 in [11], and the "running example" in [14]

$$A = \begin{pmatrix} 0 & 1 & 3 & 4 \\ 1 & 1 & 1 & 1 \end{pmatrix}.$$

As $\operatorname{vol}(A) = 4$, $\operatorname{rank}(H_A(b))$ is 4 or 5. For the vector $b = (3, 2)^\top \notin \mathscr{E}(A)$ $\operatorname{rank}(H_A(b)) = \operatorname{vol}(A) = 4$. Moreover, as $b \in \mathbb{N}_0 A$, the multiples of the A-hypergeometric polynomial $Z_A(b) = 5! x_1 x_4$ are the only Laurent solutions of the A-hypergeometric system. For the vector $b' = (2, 1)^\top \in \mathscr{E}(A)$, $\operatorname{rank}(H_A(b')) = 5$. As $b' \notin \mathbb{N}_0 A$ the A-hypergeometric polynomial is not a solution of the A-hypergeometric system. Two Laurent polynomial solutions are x_2^2/x_1 and x_4^2/x_3.

Let us summarize the above observations.

Proposition 3.1 *For a homogeneous two-row matrix A and given $b \in \mathbb{N}_0 A$, the unique polynomial solutions of the A-hypergeometric system $H_A(b)$ are constant multiples of the A-hypergeometric polynomial (3.3). Especially, the A-hypergeometric polynomial with (3.5) is a constant multiple of the associated partial Bell polynomial (2.35).*

Example 3.5 Let us consider the A-hypergeometric system with (3.5) and $r = n$. The explicit solution basis for $n = k + 2 \geq 4$ is obtained as follows. The A-hypergeometric ideal $H_A(b)$ consists of the annihilators

$$L_1 = \theta_2 + 2\theta_3 - 2, \qquad L_2 = \theta_1 + \theta_2 + \theta_3 - n + 2, \qquad L := \partial_1\partial_3 - \partial_2^2,$$

and $\mathrm{rank}(H_A(b)) = 2$ (see Example 3.3). Suppose that A-hypergeometric ideal annihilates a function $g(x_1, x_2, x_3)$. Under the annihilators L_1 and L_2, there exists a univariate function $f(y)$ such that

$$g(x_1, x_2, x_3) = x_1^{n-4}x_2^2 f(y), \qquad y := \frac{x_1 x_3}{x_2^2},$$

and the annihilator L encodes the Gauss hypergeometric equation:

$$L \bullet g(x_1, x_2, x_3) = 4x_1^{n-4}\{y(1 - 4y)f''(y) + (n - 3 + 2y)f'(y) - 2f(y)\} = 0.$$

Therefore, we have the hypergeometric polynomial

$$f(y) = {}_2F_1(-1, -1/2; n - 3; 4y) = 1 + \frac{2y}{n - 3},$$

and the A-hypergeometric polynomial is given by

$$Z_A(b; x) = \frac{x_1^{n-4}x_2^2}{2(n - 4)!}\left(1 + \frac{2y}{n - 3}\right), \tag{3.6}$$

which is $n!$ times the partial Bell polynomial $B_{n,n-2}(x..!)$. There exists another independent solution. It is a logarithmic solution obtainable by the perturbation method introduced at the end of Sect. 3.1. For a weight vector $w = (1, 0, 0)$, the reduced Gröbner basis of the toric ideal is $\{\partial_1\partial_3 - \partial_2^2\}$. Solving $v_1 v_3 = 0$ and $Av = b$, we obtain the fake exponents $v_1 = (n - 4, 2, 0) \in \mathbb{N}_0^3$ and $v_2 = (0, 2n - 6, 4 - n)$. The unique polynomial solution v_1 around the origin gives (3.6). For the other solution, we take $b' = (0, 1)^\top$ and obtain the corresponding exponents of $H_A(b + \varepsilon b')$ as

$$v_1' = v_1 + (1, 0, 0)^\top\varepsilon, \qquad v_2' = v_2 + (0, 2, -1)^\top\varepsilon.$$

The logarithmic solution is then obtained by canceling the term of order ε^{-1}:

$$\lim_{\varepsilon \to 0}\left\{\frac{1}{2\varepsilon}\frac{\phi_{v_1'}(x)}{(n - 4)!} - \frac{(n - 5)!}{(2n - 6)!}(-1)^n\phi_{v_2'}(x)\right\}, \qquad n \geq 5.$$

The result is

$$Z_A(b; x) \log y + x_2^{2n-6} x_3^{4-n} \left\{ \frac{y^{n-2}}{(n-2)!} {}_3F_2\left(\frac{3}{2}, 1, 1; n-1, 3; 4y\right) \right.$$

$$\left. - \frac{y}{(n-3)(n-3)!} - \frac{(n-5)!}{(2n-6)!}(-1)^n \sum_{i=0}^{n-5} \frac{(3-n)_i(7/2-n)_i}{(5-n)_i} \frac{(-4y)^i}{i!} \right\}. \quad (3.7)$$

This cancelation is inapplicable when $n = 4$ because the two fake exponents degenerate; however, a similar derivation for $n \geq 5$ gives the result for $n = 4$. Replacing the last term in (3.7) by $(-4y)$ gives the result.

Let us see the relationship between the A-hypergeometric system and the partial Bell polynomials explained in Sect. 2.2. The exponential generating function of the partial Bell polynomials (2.17) can be recast as

$$Z_A(b; x) = [\xi^n] \frac{\{\mathscr{W}(\xi)\}^k}{k!}, \qquad \mathscr{W}(\xi) = \sum_{i \geq 1} x_i \xi^i. \quad (3.8)$$

By an argument on the de Rham cohomology, the A-hypergeometric ideal $H_A(b)$ eliminates the A-hypergeometric integral (see Sect. 5.4 of [1]). For a homogeneous two-row matrix A and an integer-valued vector b, the integral is

$$\Phi_C(A, b; x) := \frac{1}{2\pi\sqrt{-1}} \int_C f(\xi, x)^{b_2} \xi^{-b_1-1} d\xi, \qquad f(\xi, x) := \sum_{i=1}^m x_i \xi^{a_{1i}}.$$

If the cycle C belongs to the homology group $H_1(\xi \in \mathbb{C} \backslash \{0\} | f(\xi, x) \neq 0)$, this expression is an element of the solution basis, because $H_A(b)$ annihilates $\Phi_C(A, b; x)$ by Stokes' theorem (see Theorem 5.4.2 of [1]). The expression can be regarded as a pairing of cycle C and the one-form $f^{b_2} \xi^{-b_1-1} d\xi$. For the A-hypergeometric system with (3.5) and $r = n$, let C be a cycle around the origin. The residue of the origin gives

$$\Phi_C(A, b; x) = [\xi^n] \left\{ \sum_{i=1}^{n-k+1} x_i \xi^i \right\}^k,$$

which is a constant multiple of (3.8). Here, the vector b is resonant (Condition 2.9 of [8]), and the other elements of the solution basis cannot be represented in integral form. Nevertheless, they can be obtained by perturbing the vector b, as we demonstrated in Example 3.5.

Remark 3.1 A complex-valued vector b requires twisted de Rham theory, which considers cycles of singular simplices with coefficients. As shown in [1, 8], the A-hypergeometric integral gives the full solution basis.

For later discussions, let us prepare some results for a matrix A of the form (3.4) with $i_m = m - 1$. The following assertion can be straightforwardly confirmed by computing the S-polynomials.

Proposition 3.2 *A minimal Gröbner basis of the toric ideal I_A, where the matrix A of the form (3.4) with $i_{m-1} = m - 1 \geq 2$, is*

$$G_A = \{\partial_i \partial_j - \partial_{i+1} \partial_{j-1}; 1 \leq i < j \leq m, i + 2 \leq j\}.$$

Remark 3.2 The reduced Gröbner basis can be obtained by the minimal Markov basis. The reduced Gröbner basis with $m = 4$ for the polynomial ring is given in Example 3.1.

With the aid of G_A, we obtain the standard monomials.

Proposition 3.3 ([10]) *For a matrix A of the form (3.4) with $i_{m-1} = m - 1 \geq 2$ and any vector $b \in \mathbb{C}^2$, the totality of the standard monomials of the A-hypergeometric ideal $H_A(b)$ is $\{1, \partial_i; 3 \leq i \leq m\}$.*

3.2 A-Hypergeometric Distributions

This section introduces the A-hypergeometric distributions defined by Takayama, et al. [15]. The A-hypergeometric distribution is a class of discrete exponential families, which naturally appears under the exchangeability assumption, and in multinomial sampling from discrete exponential families called log-affine models. An A-hypergeometric distribution is the conditional distribution given the sufficient statistics.

Definition 3.4 ([15]) Let A be a homogeneous $d \times m$-matrix of rank d with nonnegative integer entries, and let $b \in \mathbb{N}_0$ be a fixed vector. The A-hypergeometric distribution of the matrix A and vector b is the discrete probability mass function of the count vector $(c_1, ..., c_m)$,

$$p_{A,b}(c; x) := \frac{1}{Z_A(b; x)} \frac{x^c}{c!}, \qquad x \in \mathbb{R}^m_{>0}, \tag{3.9}$$

where $Z_A(b; x)$ is the A-hypergeometric polynomial defined in Definition 3.3, and the support is the b-fiber of matrix A, namely, $\mathscr{F}_b(A) := \{c; Ac = b\}$.

The log-likelihood is given by

$$\ell(c; \xi) = \sum_{i=1}^m (\xi_i c_i - \log c_i!) - \psi(\xi), \qquad \xi_i := \log x_i \in \mathbb{R}, \qquad i \in [m], \tag{3.10}$$

where $\psi(\xi) := \log Z_A(b; e^\xi)$. As the natural parameter space $\{\xi; Z_A(b; e^\xi) < \infty\}$ is open, the probability mass function (3.9) is regular [16, 17].

We now derive the probability mass function (3.9). Consider that among m categories, t_i is a category of the i-th observation in a sample of size n. The count vector of sample $(t_1, \ldots, t_n)$ is $(c_1, \ldots, c_m)$, $c_i := \#\{j; t_j = i\}$, where the homogeneity condition is $c_1 + \cdots + c_m = n$. The homogeneity condition fixes the total number of counts as n. For an n-exchangeable sequence $(T_1, \ldots, T_n)$ of random variables, we have

$$\mathbb{P}(C_1 = c_1, \ldots, C_m = c_m) = \frac{n!}{c_1! \cdots c_m!} \mathbb{P}(T_1 = t_1, \ldots, T_n = t_n).$$

With this expression, the count vector can be modeled as follows. Let us assume

$$\mathbb{P}(C = c) = \frac{n!}{c!} \prod_{i=1}^{m} \{p_i(\xi)\}^{c_i}, \tag{3.11}$$

where each $p_i(\xi)$ follows a discrete exponential family as

$$\log p_i(\xi) = \sum_{j=1}^{d+g} a_{ji}\xi_j - \phi_i(\xi) \;\Leftrightarrow\; p_i(\xi) = e^{-\phi_i(\xi)} \prod_{j=1}^{d+g} x_j^{a_{ji}} \tag{3.12}$$

with parameters $(\xi_1, \ldots, \xi_d, \xi_{d+1}, \ldots, \xi_{d+g})$. This model is called a *log-affine model* or a *toric model*. Substituting (3.12) into (3.11), we have the unconditional model:

$$\mathbb{P}(C = c) = \frac{n!}{c!} \exp\left\{ \sum_{j=1}^{d+g} \xi_j b_j - \sum_{i=1}^{m} \phi_i(\xi) \right\}. \tag{3.13}$$

This probability mass function is a discrete exponential family, where $(b_1, \ldots, b_{d+g})$ are sufficient statistics for the parameters $(\xi_1, \ldots, \xi_{d+g})$ satisfying

$$b_j = \sum_{i=1}^{m} a_{ji} c_i, \quad j \in [d+g].$$

The above equations for $j \in [d]$ constitute the condition $Ac = b$ in the definition of the A-hypergeometric polynomial, and the conditional distribution is

$$\mathbb{P}(C = c | AC = b) \propto \frac{1}{c!} \exp\left\{ \sum_{j=d+1}^{d+g} \xi_j b_j \right\} = \frac{x^c}{c!}, \quad \log x_i = \sum_{j=d+1}^{d+g} a_{ji}\xi_j.$$

Hence, we conclude that $p_{A,b}(c; x) = \mathbb{P}(C = c | AC = b)$.

The above parameters $(\xi_1, ..., \xi_d)$ are regarded as the nuisance parameters. A typical statistical application is the *similar test*, in which the rejection probability under the null hypothesis does not depend on the nuisance parameters. When we know the sufficient statistics for the nuisance parameters, the nuisance parameters are excluded from the conditional distribution. Hence, the conditional test becomes a similar test [18].

Example 3.6 (Poisson regression) In Sect. 1.2.2, we discussed a goodness-of-fit test in m-level univariate Poisson regression. As $C_i \sim \mathrm{Po}(\mu_i)$, where $\mu_i = \exp(\alpha + \beta i)$ with $i \in [m]$, we have

$$\mathbb{P}(C_1 = c_1, ..., C_m = c_m) = \exp\left(-\alpha m + \beta \frac{m(m+1)}{2}\right) \frac{\exp(\alpha k + \beta n)}{c!},$$

where $k = c_1 + \cdots + c_m$ and $n = 1 \cdot c_1 + \cdots + m \cdot c_m$ are the sufficient statistics. When we are interested in the fitting only, the parameters α and β are both considered as nuisance parameters. Then, given sufficient statistics, the conditional distribution is an A-hypergeometric distribution:

$$\mathbb{P}(C_1 = c_1, ..., C_m = c_m | N(C) = n, K(C) = k) = \frac{1}{Z_A(b; 1.)} \frac{1}{c!},$$

where

$$A = \begin{pmatrix} 1 & 2 & \cdots & m \\ 1 & 1 & \cdots & 1 \end{pmatrix}, \qquad b = \begin{pmatrix} n \\ k \end{pmatrix},$$

$K(C) := C_1 + \cdots + C_n$, and $N(C) := 1 \cdot C_1 + 2 \cdot C_2 + \cdots n \cdot C_n$. Here, the A-hypergeometric polynomial $Z_A(b; 1.)$ is an $n!^{-1}$ multiple of the associated signless Lah number in Example 2.18.

Example 3.7 (Two-way contingency tables) Let us consider a 2×2 contingency table.

$$\begin{array}{cc|c} n_{11} & n_{12} & n_{1\cdot} \\ n_{21} & n_{22} & n_{2\cdot} \\ \hline n_{\cdot 1} & n_{\cdot 2} & n_{\cdot\cdot} \end{array}$$

The standard model of counts is $(N_{11}, N_{12}, N_{21}, N_{22}) \sim \mathrm{Multi}(p_{11}, p_{12}, p_{21}, p_{22})$. Under the parameterization

$$\psi_1 := \log \frac{p_{12}}{p_{22}}, \qquad \psi_2 := \log \frac{p_{21}}{p_{22}}, \qquad \lambda := \log \frac{p_{11}p_{22}}{p_{12}p_{21}},$$

we have

$$\mathbb{P}(N = n) = \frac{n_{\cdot\cdot}!}{n_{11}!n_{12}!n_{21}!n_{22}!}$$
$$\times \exp\left\{n_{11}\lambda + n_{1\cdot}\psi_1 + n_{\cdot 1}\psi_2 - n_{\cdot\cdot} \log\left(1 + e^{\psi_1} + e^{\psi_2} + e^{\lambda + \psi_1 + \psi_2}\right)\right\}.$$

We are usually interested in the odds ratio $y = e^{\lambda}$, regarding ψ_1 and ψ_2 as the nuisance parameters. The total number of counts n is fixed, so $n_1.$ and $n._1$ are sufficient statistics, and the conditional distribution of the table with fixed marginal sums is free of the nuisance parameters:

$$\mathbb{P}(N_{11} = n_{11}|N_{..} = n_{..}, N_{1.} = n_{1.}, N_{.1} = n_{.1}) = \frac{1}{Z_A(b; x)} \frac{y^{n_{11}}}{n_{11}!n_{12}!n_{21}!n_{22}!}.$$

This probability mass function is the A-hypergeometric distribution with

$$A = \begin{pmatrix} 1 & 1 & 0 & 0 \\ 0 & 0 & 1 & 1 \\ 1 & 0 & 1 & 0 \\ 0 & 1 & 0 & 1 \end{pmatrix}, \quad c = \begin{pmatrix} n_{11} \\ n_{12} \\ n_{21} \\ n_{22} \end{pmatrix}, \quad b = \begin{pmatrix} n_{1.} \\ n_{2.} \\ n_{.1} \\ n_{.2} \end{pmatrix}, \quad x = \begin{pmatrix} y \\ 1 \\ 1 \\ 1 \end{pmatrix}.$$

Here, the A-hypergeometric polynomial $Z_A(b; x)$ is proportional to the Gauss hypergeometric polynomial

$$_2F_1(-n_{1.}, -n_{.1}, n - n_{1.} - n_{.1} + 1; y) = \sum_{i \geq 0} \frac{(-n_{1.})_i (-n_{.1})_i}{(n - n_{1.} - n_{.1} + 1)_i} \frac{y^i}{i!}.$$

This A-hypergeometric distribution is called the generalized hypergeometric distribution. Under the null hypothesis $y = 1$, it reduces to the hypergeometric distribution, defined for tables with independence between rows and columns. The similar test based on the conditional distribution is called Fisher's exact test. The hypergeometric series of type $(r, r + c)$ [19] is defined as

$$F(\alpha, \beta, \gamma; y) = \sum_n \frac{\prod_{i=1}^{r-1}(n_{i.} - n_{ic})_{\alpha_i} \prod_{j=1}^{c-1}(n_{.j} - n_{rj})_{\beta_j}}{(\sum_{i=1}^{r-1} \sum_{j=1}^{c-1} n_{ij})_{\gamma}} \prod_{i=1}^{r-1} \prod_{j=1}^{c-1} \frac{y_{ij}^{n_{ij}}}{n_{ij}!}, \quad (3.14)$$

where (n_{ij}) is a $(r - 1) \times (c - 1)$ matrix with nonnegative integer entries, and y is a $(r - 1) \times (c - 1)$ matrix with complex number entries. The normalizing constant of a $r \times c$ $(r, c \geq 2)$ contingency table with fixed marginal sums can be shown to be proportional to a hypergeometric polynomial of type $(r, r + c)$ with parameters

$$\alpha = (-n_{1.}, \ldots, -n_{r-1,.}), \quad \beta = (-n_{.1}, \ldots, -n_{.,c-1}), \quad \gamma = n - \sum_{i=1}^{r-1} n_{i.} - \sum_{j=1}^{c-1} n_{.j} + 1$$

$$(3.15)$$

and variables

$$y_{ij} = \frac{p_{ij} p_{rc}}{p_{ic} p_{rj}}, \quad i \in [r - 1], \ j \in [c - 1]. \quad (3.16)$$

The conditional distribution of the table with fixed marginal sums is the A-hypergeometric distribution. The probability mass function is given by

$$\frac{1}{Z_A(b, x)} \prod_{i=1}^{r-1} \prod_{j=1}^{c-1} \frac{y_{ij}^{n_{ij}}}{n_{ij}!},$$

with

$$A = \begin{pmatrix} E_r \otimes I_c' \\ I_r' \otimes E_c \end{pmatrix}, \quad c = (n_{11}, ..., n_{1c}, n_{21}, ..., n_{2c}, ..., n_{rc})^\top, \quad b = (n_{1\cdot}, ..., n_{r\cdot}, n_{\cdot 1}, ..., n_{\cdot c})^\top.$$

Here, the A-hypergeometric polynomial $Z_A(b; x)$ is proportional to the above-described $(r, r + c)$-hypergeometric polynomial.

Example 3.8 (*Exponential structures*) In Sect. 2.2 we discussed the multiplicative measure on partitions induced by the exponential structure (2.16). The probability mass function of this measure is given by

$$\mu_n(\lambda) = \frac{n!}{B_n(w)} \frac{x^c}{c!}, \quad x_i = \frac{w_i}{i!}, \quad \lambda \in \mathscr{P}_n,$$

where the size index of partition λ is given by $c_i(\lambda) = \#\{j; c_j(\lambda) = i\}$. Recall that this distribution satisfies the conditioning relation

$$\mathbb{P}(C_1 = c_1, ..., C_n = c_n | N(C) = n)$$

with $C_i \sim \mathrm{Po}(x_i \zeta^i)$ for some $\zeta > 0$. Given the length of partition (2.14), the conditional distribution can be written as

$$\mathbb{P}(C_1 = c_1, ..., C_n = c_n | N(C) = n, K(C) = k) = \frac{1}{Z_A(b; x)} \frac{x^c}{c!}, \quad \lambda \in \mathscr{P}_{n,k},$$

where

$$A = \begin{pmatrix} 1 & 2 & \cdots & n \\ 1 & 1 & \cdots & 1 \end{pmatrix}, \quad b = \begin{pmatrix} n \\ k \end{pmatrix}.$$

Hence, the conditional distribution is an A-hypergeometric distribution. Here, $n! Z_A (b; x)$ is the partial Bell polynomials $B_{n,k}(w)$ defined in (2.15). For $n = k + 2$, we have

$$Z_A(b; x) \propto {}_2F_1(-1, -1/2, n - 3; 4y) = 1 + \frac{2y}{n - 3}, \quad y := \frac{x_1 x_3}{x_2^2},$$

as shown in Example 3.5. Note that the conditional probability measure of a Gibbs partition with given length has the same expression.

A-hypergeometric polynomials (3.3) satisfy the *contiguity relation* [20]

$$\theta_i Z_A(b; x) = x_i Z_A(b - a_i; x), \qquad i \in [m], \tag{3.17}$$

where a_i is the i-th column vector of the matrix A. The moments are given by

$$\mathbb{E}(C_i) = \partial_\xi \psi(\xi) = \frac{\theta_i Z_A(b; x)}{Z_A(b; x)} = \frac{Z_A(b - a_i; x)}{Z_A(b; x)} x_i. \qquad i \in [m]. \tag{3.18}$$

The moments satisfy the polynomial constraint $A\mathbb{E}(C) = b$. Therefore, the A-hypergeometric distributions clearly comprises the algebraic exponential family defined and studied by Drton and Sullivant [21].

The following proposition generalizes Proposition 4.2 of [10] and Theorem 2.5 in [22] for the Dirichlet-multinomial distribution (4.10). It directly follows from the derivation of the A-hypergeometric distribution introduced at the beginning of Sect. 3.2, and from the Lehmann–Scheffé theorem [23].

Proposition 3.4 *For the A-hypergeometric distribution defined in Definition 3.4, the unique minimum variance unbiased estimator (UMVUE) of the joint factorial moments under the unconditional model (3.13) is given by*

$$\mathbb{E}\left[\prod_{i=1}^{n}[C_i]_{r_i} | AC = b\right] = \frac{Z_A(b - \sum_{i=1}^{m} r_i a_i; x)}{Z_A(b; x)} x^r I_{\{b - \sum_{i=1}^{m} r_i a_i \geq 0\}}.$$

If the vector $v \geq 0$, all elements of v are nonnegative.

The expression (3.18) implies that the moments are invariant under a *torus action*:

$$x_j \mapsto x_j \prod_{i=1}^{d} s_i^{a_{ij}}, \qquad j \in [m]. \tag{3.19}$$

In other words, $\xi - \xi' \in \mathrm{Im}A^\top$ have the same moment. The background of the torus action is explained in Sect. 2.3 of [1]. Takayama et al. [15] defined the *generalized odds ratio* for parameterizing the quotient space $\mathbb{R}^m/\mathrm{Im}A^\top$. This is given by

$$y_i := x^{\bar{a}_i} = \prod_{j=1}^{m} x_j^{\bar{a}_{ij}}. \qquad i \in [m - d]. \tag{3.20}$$

Here, $\bar{a}_1, ..., \bar{a}_{m-d}$ are m-dimensional row vectors and the matrix

$$\bar{A} = \begin{pmatrix} \bar{a}_1 \\ \vdots \\ \bar{a}_{m-d} \end{pmatrix},$$

called the Gale transformation of A [1, 15], satisfies $\bar{A}A^\top = 0$. For an $d \times m$ matrix A over a field, there exists the Smith normal form as

$$SAR = \begin{pmatrix} \alpha_1 & & 0 & 0 \cdots 0 \\ & \ddots & & \vdots & \vdots \\ 0 & & \alpha_d & 0 \cdots 0 \end{pmatrix},$$

where S and R are invertible $d \times d$ and $m \times m$ matrices, respectively, $\alpha_i | \alpha_{i+1}$, and $\alpha_i \neq 0$. We thus obtain $\bar{a}_i = (Re_{d+i})^\top$.

Now, parameterization with the generalized odds ratio yields the one-to-one moment map:

$$\mathbb{E}(C) : \mathbb{R}^m / \mathrm{Im}A^\top \to \mathrm{relint}(\mathrm{New}(Z_A(b; x))), \tag{3.21}$$

where the Newton polytope $\mathrm{New}(Z_A(b; x))$ is the convex hull of the support. Takayama et al. [15] (Theorem 1) established the following theorem.

Theorem 3.4 ([15]) *Let A be a $d \times m$ homogeneous matrix with nonnegative integer entries. If the affine dimension of the Newton polytope $\mathrm{New}(Z_A(b))$ is $m - d$, then the image of the moment map (3.21) agrees with the relative interior of the Newton polytope. Moreover, the moment map is one-to-one.*

Theorem 6 in [15] established an asymptotic normality for A-hypergeometric distributions for γb with $b \in \mathbb{N}A \cap \mathrm{int}(\mathbb{R}_{\geq 0}A)$ as $\gamma \to \infty$. The following asymptotic form of the A-hypergeometric polynomial is derived from the normalization constant.

Theorem 3.5 ([15]) *For an A-hypergeometric polynomial $Z_A(b; x)$, there exists a unique $\mu \in \mathbb{R}^m_{>0}$ such that $A\mu = b$ and $y = \mu^{\bar{A}}$. Moreover,*

$$Z_A(\gamma b; x) \sim \frac{(x^\mu)^\gamma}{\Gamma(\gamma \mu + 1)} \frac{(2\pi \gamma)^{m-d}}{\det(\bar{A}M^{-1}\bar{A}^\top)^{1/2}}, \qquad \gamma \to \infty,$$

where $M = \mathrm{diag}(\mu)$ and $\Gamma(\gamma \mu + 1) = \prod_{i=1}^m \Gamma(\gamma \mu_i + 1)$.

Remark 3.3 The derivation of μ must be explained. Suppose that a count vector c follows independent Poisson distributions with log-affine models

$$\mathbb{P}(C = c) = \exp(-1 \cdot p)\frac{p^c}{c!}, \qquad \log p(\xi) = A^\top \xi + \bar{A}\bar{A}^\top (\bar{A}\bar{A}^\top)^{-1} \log y.$$

The maximum likelihood estimator (MLE) $\hat{\xi}(y)$ can be numerically evaluated by the iterative proportional scaling procedure [24]. Here, $\mu = p(\hat{\xi}(y))$ is the unique solution of $A\mu = Ac$ and $y = \mu^{\bar{A}}$.

3.2.1 Maximum Likelihood Estimation with a Two-Row Matrix

In this subsection, we investigate the maximum likelihood estimation of a specific class of A-hypergeometric distributions, namely, the A-hypergeometric distribution with (3.5) and $r = n$. The maximum likelihood estimation of another class of A-hypergeometric distributions, the A-hypergeometric distribution of matrix A of $2 \times c$ contingency table, was discussed in Ogawa's thesis [25].

Under the torus action (3.19) the A-hypergeometric polynomial transforms as

$$Z_A(b; s_1^{-1} s_2 x.) = s_1^{n-k} s_2^k Z_A(b; x).$$

This is a known property of partial Bell polynomials (p. 135 in [13]). The generalized odds ratios (3.20) becomes

$$y_i = \frac{x_1^i x_{i+2}}{x_2^{i+1}}, \qquad i \in [n-k-1]. \tag{3.22}$$

Under the torus action with $s_1 = x_2^{-1} x_1$ and $s_2 = x_1^{-1}$ in (3.19), we can set $x = (1, 1, y_1, ..., y_{n-k-1})$ without loss of generality.

The convex hull of all exponent vectors appearing in the normally ordered expression in the homogenized Weyl algebra is called the *Newton polytope* (see Sect. 2.1 of [1]). For the A-hypergeometric polynomial with (3.5) and $r = n$, the support comprises the integer partitions $\mathscr{P}_{n,k} := \{\lambda; \lambda \vdash n, l(\lambda) = k\}$. For an integer partition $\lambda \in \mathscr{P}_{n,k}$, the size index $(c_1, ..., c_{n-k+1}) \in \mathbb{N}_0^{n-k+1}$ is an integer point and the convex hull of $\mathscr{P}_{n,k}$ is the Newton polytope $\mathrm{New}(Z_A(b; x))$. To aid the following discussion, we also define the *partition polytope*. For an integer partition $\lambda \in \mathscr{P}_n = \cup_{k=1}^n \mathscr{P}_{n,k}$, the size index $(c_1, ..., c_n) \in \mathbb{N}_0^n$ is an integer point, and the partition polytope P_n is the convex hull of $\mathscr{P}_n$. The properties of partition polytopes were studied by Shlyk [26]. We are interested in a fundamental property of partition polytopes, namely, that P_n is a pyramid with apex e_n, where e_n is the single vertex of P_n with $c_n > 0$, and the other integer points comprising the base of the pyramid locate in the hyperplane $c_n = 0$. P_n is the convex hull of e_n and the base, meaning that all vertices of P_n lie on the faces of P_n. Precisely, the base contains the polytope P_{n-1} translated by 1 along the c_1 axis and embedded into $\mathbb{N}_0^n$. Identifying P_{n-1} and its image under the translation

$$\varphi_1 : (c_1, c_2, ..., c_{n-1}) \mapsto (c_1 + 1, c_2, ..., c_{n-1}, 0)$$

we can regard P_{n-1} as a part of P_n. In this convention, the partition polytopes constitute an embedded chain $P_1 \subset P_2 \subset \cdots$. Theorem 2 of [26] gives the following property of embedded chains.

Theorem 3.6 ([26]) *All vertices of P_n, except e_n, comprise the φ_i-image of the vertices of some preceding polytopes P_{n-i}, $i \in \lfloor n/2 \rfloor$, where*

$$\varphi_i : (c_1, ..., c_{n-i}) \mapsto (c_1, ..., c_{i-1}, c_i + 1, c_{i+1}, ..., c_{n-i}, 0, ...)$$

If $v \neq e_n$ is a vertex of P_n and $i = \min(j; c_j > 0)$, then $v = \varphi_i(u)$ for some vertex u of P_{n-i}.

Because the gradient of the log-likelihood (3.10) is $\partial_i \ell = c_i - \eta_i$, finding the MLE is equivalent to finding the inverse image of the moment map (3.21). The following theorem directly follows from the pyramid property of partition polytopes.

Theorem 3.7 ([10]) *For the likelihood given by the A-hypergeometric distribution of the matrix A and vector b in (3.5) with $r = n$, the MLE does not exist with probability one.*

Proof Because the probability mass function (3.9) is regular, the MLE exists if and only if the point c represented by sufficient statistics locates in the interior of the convex hull of the support (Theorem 5.5 of [16] and Corollary 9.6 of [17]). Let us show that c never locates in the relative interior of the Newton polytope $\text{New}(Z_A(b; x))$. If $n \geq k \geq n/2$, there is a one-to-one affine map between a partition in $\mathscr{P}_{n-k}$ and that in $\mathscr{P}_{n,k}$:

$$\mathscr{P}_{n-k} \ni (c_1, ..., c_{n-k}, 0) \mapsto \left(k - \sum_{i=1}^{n-k} c_i, c_1, ..., c_{n-k} \right) \in \mathscr{P}_{n,k}.$$

This map is easily confirmed on a Young tableau; if we erase the rightmost column of a partition in $\mathscr{P}_{n-k}$, we obtain a partition in $\mathscr{P}_{n-k}$. All vertices of the partition polytope P_{n-k} locate on the faces of P_{n-k}. Since any vertex on the face of P_{n-k} is mapped to a vertex on the face of $\text{New}(Z_A(b; x))$ by the one-to-one affine map, all vertices of $\text{New}(Z_A(b; x))$ locate on the faces of $\text{New}(Z_A(b; x))$. If $2 \leq k < n/2$, the modified map

$$\tilde{\mathscr{P}}_{n-k} \ni \left\{ (c_1, ..., c_{n-k}, 0) : \sum_{i=1}^{n-k} c_i \leq k \right\} \mapsto \left(k - \sum_{i=1}^{n-k} c_i, c_1, ..., c_{n-k} \right) \in \mathscr{P}_{n,k}$$

is one-to-one, where $\tilde{\mathscr{P}}_{n-k}$ is the collection of all integer partitions of $n - k$ with $\sum_{i=1}^{n-k} c_i \leq k$. Using this map, the assertion can be shown as done for $n \geq k \geq n/2$. $\qquad\square$

Remark 3.4 This assertion has an analogy in the theory of exponential families. If the sample size is one, the MLEs of the beta and gamma distributions do not exist with probability one, because the sufficient statistics are on the boundary of the parameter space (see Example 5.6 in [16] and Example 9.8 in [17]). The size index of the *A*-hypergeometric distribution can be regarded as a multivariate sample of size one.

From a statistical viewpoint, the nonexistence of the MLE shown in Theorem 3.7 arises from overparameterization of the model. The MLE can be recovered in two ways: by reducing the parameter space using the curved exponential family (as discussed later in Sect. 5.2), or by assuming a sample consisting of multiple count vectors, each of which follows the A-hypergeometric distribution independently. If the size indices $\{c^{(1)}, ..., c^{(N)}\}$ are i.i.d., the log-likelihood becomes $N \left\{ \sum_i \xi_i \bar{c}_i - \psi(\xi) \right\}$ with $\bar{c}_i := N^{-1} \sum_{j=1}^{N} c_i^{(j)}$. When $N \geq 2$, $\bar{c}$ can enter the relative interior of the Newton polytope, so the MLE can exist. The identifiability is given by the following corollary of Theorem 3.4. The proof relies on the affine dimension of P_n being equal to $n - 1$ (Theorem 1 of [26]), which follows from the discussion at the beginning of this subsection.

Corollary 3.1 ([10]) *For the A-hypergeometric distribution of the matrix A and vector b in (3.5) with $r = n$, the image of the moment map (3.21) agrees with the relative interior of the Newton polytope* New($Z_A(b; x)$). *Moreover, the moment map is one-to-one.*

Example 3.9 In Example 3.5, we considered an A-hypergeometric system for $n = k + 2 \geq 4$. The Newton polytope is the finite open interval between the two possible observations $(n - 4, 2, 0)^\top$ or $(n - 3, 0, 1)^\top$. The image of the moment map is

$$\begin{pmatrix} \eta_1 \\ \eta_2 \\ \eta_3 \end{pmatrix} = \begin{pmatrix} n - 4 \\ 2 \\ 0 \end{pmatrix} + \left(1 + \frac{n - 3}{2y_1}\right)^{-1} \begin{pmatrix} 1 \\ -2 \\ 1 \end{pmatrix}, \qquad y_1 \in \mathbb{R}_{>0},$$

where y_1 is the generalized odds ratio defined in (3.20). If $N = 1$, there are two possible observations $(n - 3, 0, 1)$ and $(n - 4, 2, 0)$, with likelihoods of $(1 + (n - 3)/(2y_1))^{-1}$ and $(1 + 2y_1/(n - 3))^{-1}$, respectively. The MLE does not exist in either of these cases. Let us now assume $N \geq 2$ and let the numbers of observations of $(n - 3, 0, 1)^\top$ and $(n - 4, 2, 0)^\top$ be N_1 and $N_2 = N - N_1$, respectively. Here, $N_1 \sim$ Bin($N, \eta_3(y)$). The MLE of y_1 exists for a sample with

$$\bar{c} = \left(\frac{(n - 3)N_1 + (n - 4)N_2}{N}, \frac{2N_2}{N}, \frac{N_1}{N} \right), \qquad 0 < N_1 < N.$$

The MLE of y_1 is the unique solution of $\eta(y_1) = \bar{c}$, and $\hat{y}_1 = N_1/N_2 \times (n - 3)/2$. The MLE is consistent in $N \to \infty$ and the asymptotic variance is

$$\text{Var}(\log \hat{y}_1) \sim \frac{\{y_1 + (n - 3)/2\}^2}{Ny_1(n - 3)/2}.$$

Remarkably, the asymptotic variance increases linearly with sample size n when n is large.

3.3 Holonomic Gradient Methods

Hypergeometric functions can be evaluated through recurrence relations satisfied by hypergeometric functions. This section discusses the evaluation of A-hypergeometric polynomials by various methods using recurrence relations.

Holonomic gradient methods (HGMs) are methods to solve holonomic system while avoiding combinatorial enumerations. Various applications can be found in [27]. In numerical evaluations of A-hypergeometric polynomials (3.3), we wish to avoid enumeration of $Ac = b$. The first step in HGM construction finds a system of partial differential equations for the holonomic ideal I:

$$\theta_i \bullet Q = P_i Q, \qquad i \in [\text{rank}(I)], \tag{3.23}$$

where the vector Q constitutes the totality of the standard monomials (see Sect. 3.1.1). Such a system is called a *Pfaffian system*. A Pfaffian system follows from a unique relation among standard monomials

$$\partial_i \bullet \partial_{p(j)} - \sum_{l=1}^{\text{rank}(I)} p_{ijl}(x) \bullet \partial_{p(l)} \in RI, \qquad i \in [m], \ j \in [\text{rank}(I)],$$

where the coefficients $p_{ijl}(x)$ are rational functions which are obtained by Gröbner basis normal forms with respect to a Gröbner basis. In principle, a Gröbner basis is obtained through the Buchberger's algorithm and then the normal forms can be computed (Theorem 1.4.22 of [1]; see also Sect. 6.2 of [4]). However, the procedure may be unrealistic in practice, because the computational cost grows rapidly with the rank. In practice, explicit forms of a Pfaffian system and/or efficient methods for their computation are needed.

The original HGM was formulated as follows [28]. Suppose that the value of Q at a point x_0 is known. If we know a Pfaffian system, we can compute the difference

$$Q(x + h) - Q(x) \approx \sum_i \frac{h_i}{x_i} \theta_i \bullet Q = \sum_i \frac{h_i}{x_i} P_i Q$$

for small h_i. A numerical integration method for difference equations, such as the Runge–Kutta method, provides the approximate value of $Q(x)$ at an arbitrary point x. Ohara and Takayama [29] developed another type of HGM for A-hypergeometric polynomials. Using the contiguity relation (3.17), they constructed a discrete analog of the above-described HGM. Substituting (3.17) into a Pfaffian system (3.23) eliminates the derivative, and difference equations for Q are available. The Pfaffian system is computed along a line $b - iv, i \in \mathbb{N}$, with a direction vector v, which should contain a point at which the value of Q is known. Ohara and Takayama [29] referred to these methods *difference HGMs*, and discussed the construction of a Pfaffian system for a general matrix A. The author is aware of two classes of A-hypergeometric polynomial with explicit forms of Pfaffian systems. One class constitutes the A-hypergeometric polynomials of two-row matrix A; the other comprises hypergeometric functions of type $(r, r + c)$, whose polynomial versions are the normalizing constants of the two-way contingency tables with fixed marginal sums (Example 3.7). In the following, we present a difference HGM of a two-row matrix [10], then briefly introduce a difference HGM for hypergeometric polynomials of type $(r, r + c)$, obtained by Goto and Matsumoto [30].

Let us consider the A-hypergeometric system with (3.5), $r = n$, and $n \geq k + 2$ ≥ 4. The cases of $n = k$ and $n = k + 1$ are trivial. As shown in Proposition 3.1, the A-hypergeometric polynomial is a constant multiple of the partial Bell polynomial. Therefore, A-hypergeometric polynomial can be numerically evaluated by the recurrence relation in Proposition 2.3. Below we introduce difference HGM as an alternative numerical method for evaluating A-hypergeometric polynomials.

According to Proposition 3.3, the Pfaffian system (3.23) should be given as

$$\theta_i \bullet Q(b; x) = P_i(b; x)Q(b; x), \qquad i \in [n - k + 1], \tag{3.24}$$

where

$$Q(b; x) := (1, \theta_3, ..., \theta_{n-k+1})^\top \bullet Z_A(b; x).$$

Applying annihilators of higher order differential operators, the contiguity relation (3.17), and the recurrence relation in Proposition 2.3, we obtain the following result.

Lemma 3.2 ([10]) *A Pfaffian system (3.24) for an A-hypergeometric polynomial of the matrix A and vector b in (3.5) with $r = n$ and $n \geq k + 2 \geq 4$ is given by*

$$(P_i)_{lm} = \delta_{l,1}(P_i)_{1m} + \delta_{l,m}\delta_{l,i-1}1_{\{i \geq 3\}} + \{(\tilde{P}_i^{(n)})^{-1}\}_{l-1,m-i}1_{\{2 \leq l \leq n-k-i+1, i+1 \leq m \leq n-k\}}$$

for $l, m \in [n - k]$ and $i \in [n - k + 1]$, where

$$(P_1)_{1\cdot} = (2k - n, 1, 2, ..., n - k - 1), \qquad (P_2)_{1\cdot} = (n - k, -2, -3, ..., -n + k),$$
$$(P_i)_{1j} = \delta_{i,j+1}, \qquad 3 \leq i \leq n - k + 1, \ 1 \leq j \leq n - k,$$

and $\tilde{P}_i$ are upper triangular matrices with elements

$$(\tilde{P}_i^{(n)})_{lm} = \frac{m - l + 1}{n - i - l - 1} \frac{x_{m-l+1}x_{i+l+1}}{x_{m+2}x_i}, \qquad 1 \leq l \leq m \leq n - k - i.$$

In the above expressions, P_i denotes $P_i(b; x)$.

The Pfaffian system in Lemma 3.2 holds only for A-hypergeometric polynomials and is not satisfied by the general solution of the A-hypergeometric system. This is because the Pfaffian system was derived from properties specific to A-hypergeometric polynomials.

Example 3.10 As a continuation of Example 3.5, we now consider the case of $n = k + 2$. The reduced Gröbner basis is

$$\{x_1\partial_1 - x_3\partial_3 - n + 4, x_2\partial_2 + 2x_3\partial_3 - 2,$$
$$(x_2^2 - 4x_1x_3)x_3^2\partial_3^2 + (n-3)x_2^2x_3\partial_3 - x_1x_2x_3\partial_2\}.$$

The totality of the standard monomials is $\{1, \partial_3\}$, as we have seen in Proposition 3.3. The matrices in the Pfaffian system (3.24) are

$$P_1 = \begin{pmatrix} n-4 & 1 \\ \frac{2y}{1-4y} & \frac{(10-4n)y}{1-4y} \end{pmatrix}, \quad P_2 = \begin{pmatrix} 2 & -2 \\ \frac{-4y}{1-4y} & \frac{4y+2(n-3)}{1-4y} \end{pmatrix},$$
$$P_3 = \begin{pmatrix} 0 & 1 \\ \frac{2y}{1-4y} & \frac{-6y-(n-4)}{1-4y} \end{pmatrix}. \tag{3.25}$$

These matrices are singular at the singular locus $y = 1/4$. A linear combination of the solution basis (3.6) and (3.7) satisfies the Pfaffian system (3.24) with the matrices (3.25). Meanwhile, the A-hypergeometric polynomial (3.6) satisfies the Pfaffian system with the matrices given in Lemma 3.2:

$$P_1 = \begin{pmatrix} n-4 & 1 \\ 0 & n-3 \end{pmatrix}, \quad P_2 = \begin{pmatrix} 2 & -2 \\ 0 & 0 \end{pmatrix}, \quad P_3 = \begin{pmatrix} 0 & 1 \\ 0 & 1 \end{pmatrix}.$$

However, the series (3.7) does not satisfy this Pfaffian system.

Let us obtain a difference HGM for A-hypergeometric polynomials of matrix A and vector b in (3.5) with $n \geq k + 2 \geq 4$. Substituting (3.17) into the Pfaffian system (3.24), the difference equation is given by

$$x_iQ(b - a_i; x) = P_i(b; x)Q(b; x), \quad i = 1, 2. \tag{3.26}$$

For simplicity and intuitiveness, we introduce the following notations:

$$Z_{ij} := Z_A((i-j,j)^\top; x), \quad Q_{ij} := Q((i-j,j)^\top; x), \quad P_i^{(jl)} := P_i((j-l,l)^\top; x).$$

Especially, we have $Z_{nk} = Z_A(b; x)$, $Q_{nk} = Q(b; x)$, and $P_i^{(nk)} = P_i(b; x)$. For $2 \leq k < n/2$, we then have

$$Q_{nk} = x_1^{k-1} \prod_{i=0}^{k-2} \left(P_1^{(n-i,k-i)}\right)^{-1} Q_{n-k+1,1}, \tag{3.27}$$

where $(Q_{n-k+1,1})_i = (\delta_{i,1} + \delta_{i,n-k})x_{n-k+1}$ and

$$(P_1^{(ij)})^{-1} = \frac{1}{2j-i}\begin{pmatrix} 1 & -1 & -2 & \cdots & -(i-j-1) \\ 0 & & (2j-i)E_{i-j-1} & \end{pmatrix}\begin{pmatrix} 1 & 0 \\ 0 & \tilde{P}_1^{(i)} \end{pmatrix}.$$

This method fails for $n/2 \le k \le n - 2$, because $P_1^{(2j,j)}$ is singular. However, the case $n/2 \le k \le n - 2$ can be solved by the following algorithm.

Algorithm 3.2 ([10]) Compute the solution basis of the A-hypergeometric polynomial $Q(b; x)$ in (3.24) for an A-hypergeometric polynomial of matrix A and vector b in (3.5) with $r = n$. If $2 \le k < n/2$, apply (3.27). If $n/2 \le k \le n - 2$,

1. Compute $Z_{2i+1,i}$, $i = 2, 3, \ldots, n - k - 1$ using (3.27).
2. Set $Q_{4,2} = (x_1 x_3 + x_2^2/2, x_1 x_3)^\top$ and $i = 3$.
3. Compute

$$Q_{2i,i} = \frac{1}{i} \begin{pmatrix} 2x_1 & x_2 & -1 & -2 & \cdots & -(i-2) \\ ix_1 & 0 & -2i & -3i & \cdots & -i(i-1) \\ 0 & 0 & & & & E_{i-2} \end{pmatrix} \begin{pmatrix} E_2 & 0 \\ 0 & x_2 \tilde{P}_2^{(2i)} \end{pmatrix} \begin{pmatrix} Z_{2i-1,i-1} \\ Q_{2(i-1),i-1} \end{pmatrix}.$$

4. If $i < n - k$, increment i to $i + 1$ and go to Step 2.
5. Else compute

$$Q_{n,k} = x_1^{2k-n} \prod_{i=0}^{2k-n-1} \left(P_1^{(n-i,k-i)} \right)^{-1} Q_{2(n-k),n-k}.$$

Let us consider the computational cost of the difference HGM (Algorithm 3.2). The cases $2 \le k < n/2$ and $n/2 \le k \le n - 2$ require $O((n - k)^2 k)$ steps and $O((n - k)^4 + (n - k)^2(2k - n))$ steps, respectively. Simple use of the recurrence relation in Proposition 2.3 requires $O((n - k)^2 k)$ steps. Roughly speaking, the computational effort increases with rank $n - k$, and the number of steps are roughly comparable in the difference HGM and the simple recurrence. However, as shown in the following example, the difference HGM requires much less computational time than the simple recurrence, possibly because the latter involves huge numbers.

Example 3.11 (*Generalized factorial coefficients*) The generalized factorial coefficients introduced in Sect. 2.2 are given by the A-hypergeometric polynomials:

$$C(n, k; \alpha) = (-1)^n (-\alpha)^k n! Z_A((n - k, k)^\top; (1 - \alpha)_{\cdot -1}/\cdot!).$$

Note that the exact values at $\alpha = 1/2$ (2.27) are known. Table 3.1 displays the numerical results at $\alpha = 1/2$ calculated by the simple use of the recurrence relation in Proposition 2.3, and by the difference HGM (Algorithm 3.2). For comparison, the exact values obtained by (2.27) and the asymptotic values obtained by the asymptotic formula in Lemma 2.1 are also shown. All computations were performed on rational numbers and were implemented by Risa/Asir, Version 20160405 [31] on an Intel Xeon CPU E5-2680 2.80 GHz with 32 Gb memory. The computational time was consistently smaller in the difference HGM than in the simple recurrence. In the table entries containing dashes, the results were unavailable because of memory overrun and computational times exceeding one hour.

Table 3.1 Results of numerical evaluations of generalized factorial coefficients with $\alpha = 1/2$. The top, middle, and bottom are the results for $k = 4n/5$, $k = n/2$, and $k = n/5$, respectively. The exact values and the asymptotic values are also shown. Time is in seconds. Dashes indicate that the results were unavailable

n	50	100	200	400
Exact				
$\log Z$	−99.54305	−250.06708	−605.88178	−1427.7256
Asymptotic				
$\log Z$	−99.29945	−249.82778	−605.64463	−1427.4895
Recursion				
$\log Z$	−99.54305	−250.06708	−605.88178	−
Time	0.082989	1.77773	771.334	−
HGM				
$\log Z$	−99.54305	−250.06708	−605.88178	−1427.7256
Time	0.010998	0.148977	4.19437	153.563
n	50	100	200	400
Exact				
$\log Z$	−48.350547	−126.08813	−315.53294	−763.04682
Asymptotic				
$\log Z$	−49.039805	−126.77934	−316.22511	−763.73948
Recursion				
$\log Z$	−48.350547	−126.08813	−	−
Time	1.28181	1130.71	−	−
HGM				
$\log Z$	−48.350547	−126.08813	−315.53294	−763.04682
Time	0.231966	9.56455	414.03	17170.7
n	50	100	200	400
Exact				
$\log Z$	−13.399179	−34.600777	−90.182133	−228.38177
Asymptotic				
$\log Z$	−15.240772	−36.443802	−92.025903	−230.22597
Recursion				
$\log Z$	−13.399179	−	−	−
Time	5.99109	−	−	−
HGM				
$\log Z$	−13.399179	−34.600777	−90.182133	−228.38177
Time	0.082984	1.87172	41.6757	894.863

Remark 3.5 Example 6.4 of [10] also evaluated generalized factorial coefficients by the difference HGM. Although the cases examined in Example 3.11 were more computationally demanding than those in Example 6.4 of [10], they were computed by the difference HGM within a reasonable time (less than 10 seconds for $n \leq 100$). More computational time is needed for $n > 100$, but the errors in the asymptotic values for $n = 100$ were below 1%. Therefore, when n exceeds 100, we can use the asymptotic values rather than compute the difference HGM. Example 6.4 of [10] and Example 3.11 were also performed in difference arithmetics basis, whereas in Example 6.4 of [10] was computed in floating-point arithmetic, Example 3.11 was computed in rational numbers. The numerical errors in the results of the difference HGM, observed in Example 6.4 of [10], are attributable to the floating-point arithmetic [32].

Goto and Matsumoto [30] studied the twisted cohomology group of hypergeometric functions of type $(r, r + c)$. The normalizing constant of the two-way contingency tables with fixed marginal sums is the hypergeometric polynomial for the parameter vector (3.15). For a generic parameter vector, the A-hypergeometric integral aided by the twisted de Rham theory provides a full solution basis (see Remark 3.1 and the preceding discussion). Goto and Matsumoto [30] obtained the Pfaffian system and contiguity relations as matrix representations of some linear maps on the twisted cohomology group, thus generalizing their results on Lauricella's F_D (with $r = 2$) in [33, 34]. In Sect. 7 of [30], they applied the difference HGM to the normalizing constant of the two-way contingency tables with fixed marginal sums. Goto et al. [35] and Tachibana et al. [36] discussed the implementation and several devices for the efficient computations.

The rank of the twisted cohomology group is given by [19]

$$\binom{r + c - 2}{r - 1}.$$

Let $\eta = (\alpha, \beta, \gamma)$ be the parameter vector of the hypergeometric function of type $(r, r + c)$ in (3.14). Analogously to the vector Q in (3.23), Goto and Matsumoto [30] constructed a vector of hypergeometric integrals, $\Phi(\eta; x)$, which they called the Gauss–Manin vector. They also defined the shifted parameter $\eta^{(i)}$, an analog of $b - a_i$ in the contiguity relation (3.17), and determined a cohomology group by an operator of the form $\nabla^{(i)} = d_\xi + \omega^{(i)} \wedge, i \in [r + c - 1]$, where $\omega^{(i)}$ is determined by the shifted parameter $\eta^{(i)}$. They then considered the linear map between these operators. A hypergeometric integral is given by pairing of an element of the cohomology group and a twisted cycle. Their obtained contiguity relation

$$\Phi(\eta^{(i)}; x) = U_i(\eta; x)\Phi(\eta; x)$$

is analogous to (3.26). The explicit form of the matrix $U_i(\eta; x)$ contains intersection numbers between pair of elements of cohomology groups (Theorem 5.3 of [30]). Section 7 of [30] provides an example of a 3×3 contingency table. The Gauss–

Manin vector of the hypergeometric polynomials are given by

$$\left(F, \frac{x_{12}}{n_{2.}} \frac{\partial F}{\partial x_{21}}, \frac{x_{22}}{n_{3.}} \frac{\partial F}{\partial x_{22}}, -\frac{x_{11}}{n_{2.}} \frac{\partial F}{\partial x_{11}}, -\frac{x_{12}}{n_{3.}} \frac{\partial F}{\partial x_{12}}, \frac{x_{11}x_{22} - x_{12}x_{21}}{n_{2.}n_{3.}} \frac{\partial^2 F}{\partial x_{11} \partial x_{22}} \right),$$

where $x_{ij} = p_{i,j+1}p_{r1}/(p_{i1}p_{r,j+1})$. A 2×2 contingency table is exemplified in Tachibana et al. [35, 36].

References

1. Saito, M., Sturmfels, B., Takayama, N.: Gröbner Deformations of Hypergeometric Differential Equations. Springer, Berlin (2010)
2. Sturmfels, B.: Gröbner Bases and Convex Polytopes. University Lecture Notes, vol. 8. American Mathematical Socitey, Providence (1996)
3. Cox, D., Little, L., O'Shea, D.: Ideals, Varieties, and Algorithms, 3rd edn. Springer, New York (2007)
4. Hibi, T. (ed.): Gröbner Bases: Statistics and Software Systems. Springer, Tokyo (2013)
5. Pistone, G., Riccomagno, E., Wynn, H.P.: Algebraic Statistics. Computational Commutative Algebra in Statistics. Chapman & Hall, Boca Raton (2000)
6. Drton, M., Sturmfels, B., Sullivant, S.: Lectures on Algebraic Statistics. Birkhäuser, Basel (2009)
7. Aoki, S., Hara, H., Takemura, A.: Markov Bases in Algebraic Statistics. Springer, New York (2012)
8. Gel'fand, I.M., Zelevinsky, A.V., Kapranov, M.M.: Generalized Euler Integrals and A-hypergeometric functions. Adv. Math. **84**, 255–271 (1990)
9. Forsyth, A.R.: A Treatise on Differential Equations, 6th edn. Macmillan, New York (1956)
10. Mano, S.: Partition structure and the A-hypergeometric distribution associated with the rational normal curve. Electron. J. Stat. **11**, 4452–4487 (2017)
11. Cattani, E., D'Andrea, C., Dickenstein, A.: Rational solutions of the A-hypergeometric system associated with a monomial curve. Duke Math. J. **99**, 179–207 (1999)
12. Hartshorne, R. Algebraic Geometry. Graduate Texts in Mathematics, vol. 52. Springer, New York (1977)
13. Comtet, L.: Advanced Combinatorics. Ridel, Dordrecht (1974)
14. Strumfels, B., and Takayama, N.: Gröbner bases and hypergeometric functions. In: Buchberger, B., Winkler, F. (eds.) Gröbner bases and Applications, Proceedings of Conference 33 Years of Gröbner bases. London Mathematical Society Lecture Note Series, vol. 251, pp. 246–258. Cambridge University Press, Cambridge (1998)
15. Takayama, N., Kuriki, S., Takemura, A.: A-hypergeometric distributions and Newton polytopes. Adv. Appl. Math. **99**, 109–133 (2018)
16. Brown, L.D.: Fundamentals of Statistical Exponential Families with Applications in Statistical Decision Theory. Institute of Mathematical Statistics, Hayward (1986)
17. Barndorff-Nielsen, O.E.: Information and Exponential Families in Statistical Theory. Wiley, Chichester (2014)
18. Lehmann, E.L., Romano, J.P.: Testing Statistical Hypothesis, 3rd edn. Springer, New York (2005)
19. Aomoto, K., Kita, M.: Theory of Hypergeometric Functions. Springer, New York (2011)
20. Saito, M., Sturmfels, B., Takayama, N.: Hypergeometric polynomials and integer programming. Compos. Math. **115**, 185–204 (1999)
21. Drton, M., Sullivant, S.: Algebraic statistical models. Stat. Sinica **17**, 1273–1297 (2007)
22. Keener, R., Rothman, E., Starr, N.: Distribution of partitions. Ann. Stat. **15**, 1466–1481 (1978)

23. Lehmann, E.L., Casella, G.: Theory of Point Estimation, 2nd edn. Springer, New York (1998)
24. Darroch, J.N., Ratcliff, D.: Generalized iterative scaling for log-linear models. Ann. Math. Stat. **43**, 1470–1480 (1972)
25. Ogawa, M.: Algebraic statistical methods for conditional inference of discrete statistical models. Dissertation, University of Tokyo, Tokyo (2015)
26. Shlyk, V.A.: Polytopes of partitions of numbers. Eur. J. Comb. **26**, 1139–1153 (2005)
27. Takayama, N.: References for the Holonomic Gradient Method (HGM) and the Holonomic Gradient Descent Method (HGD). http://www.math.kobe-u.ac.jp/OpenXM/Math/hgm/ref-hgm. html
28. Nakayama, H., Nishiyama, K., Noro, M., Ohara, K., Sei, T., Takayama, N., Takemura, A.: Holonomic gradient descent and its application to the Fisher-Bingham integral. Adv. Appl. Math. **47**, 639–658 (2011)
29. Ohara, K., Takayama, N.: Pfaffian systems of *A*-hypergeometric systems II–holonomic gradient method. arXiv: 1505.02947
30. Goto, Y., Matsumoto, K.: Pfaffian equations and contiguity relations of the hypergeometric function of type $(k + 1, k + n + 2)$ and their applications. arXiv: 1602.01637. to appear in Funkcial. Ekvac
31. Risa/Asir (Kobe distribution) Download Page. http://www.math.kobe-u.ac.jp/Asir/asir.html
32. Tachibana, Y., Takayama, N.: Private communication
33. Matsumoto, K.: Monodoromy and Pfaffian of Lauricella's F_D in terms of the intersection forms of twisted (co)homology groups. Kyushu J. Math. **67**, 367–387 (2013)
34. Goto, Y.: Contiguity relations of Lauricella's F_D revisited. Tohoku Math. J. **69**, 287–304 (2017)
35. Goto, Y., Tachibana, Y., Takayama, N.: Implementation of difference holonomic gradient methods for two-way contingency table. Computer Algebra and Related Topics. RIMS Kôkyûroku 2054, 11 (2016) in Japanese
36. Tachibana, Y., Goto, Y., Koyama, T., Takayama, N.: Holonomic gradient method for two way contingency tables. arXiv: 1803.04170

Chapter 4
Dirichlet Processes

Abstract After preparing some basic facts in probability theory, this chapter introduces de Finetti's representation theorem. We, then, introduce the Dirichlet process and the Poisson–Dirichlet distribution, which are closely related to exchangeability. We discuss two constructions of the Dirichlet process: one based on the normalized gamma process, the other using the stick breaking. The relationship between these two constructions is revealed in terms of biased permutations. The sequential sampling scheme is known as Blackwell–MacQueen's urn scheme. A sample from the Dirichlet process follows the Ewens sampling formula, which was encountered as a measure on partitions in Chap. 2. The Ewens sampling formula is an example of exchangeable partition probability function. The Dirichlet process possesses several nice properties for statistical applications. Therefore, the Dirichlet process has been used as a fundamental prior process in Bayesian nonparametrics. We will discuss some prior processes as generalizations of the Dirichlet process. Several prior processes naturally appear in connection with infinite exchangeable Gibbs partitions introduced in Chap. 2.

Keywords Blackwell–MacQueen's urn · De Finetti's representation theorem
Dirichlet process · Exchangeability · Poisson–Dirichlet distribution
Poisson–Kingman distribution · Prediction rule · Subordinator · Gibbs partition
Normalized random measure with independent increments

4.1 Preliminaries

Let us prepare some basic concepts and definitions in probability theory, which will be used in later discussion. The details are given in standard textbooks, such as [1–3]. The reader is referred to [4, 5] in Japanese.

Consider a probability space $(\Omega, \mathscr{F}, P)$. Suppose that a real-valued function $X = X(\omega)$, $\omega \in \Omega$ is measurable under a map $X : (\Omega, \mathscr{F}) \to (\mathbb{R}, \mathscr{B}(\mathbb{R}))$; that is, X satisfies

$$\forall A \in \mathscr{B}(\mathbb{R}), \qquad X^{-1}(A) = \{\omega \in \Omega; X(\omega) \in A\} \in \mathscr{F}.$$

© The Author(s) 2018
S. Mano, *Partitions, Hypergeometric Systems, and Dirichlet Processes in Statistics*,
JSS Research Series in Statistics, https://doi.org/10.1007/978-4-431-55888-0_4

Then, X is said to be $\mathscr{F}$-measurable and is called a *random variable*. Now consider a sub-σ field $\mathscr{G} \subset \mathscr{F}$. By the Radon–Nikodym theorem, for an $\mathscr{F}$-measurable and P-integrable random variable X, there exists a unique $\mathscr{G}$-measurable and P-integrable function Y with the following property:

$$\forall B \in \mathscr{G} \Rightarrow \mathbb{E}(X, B) = \mathbb{E}(Y, B).$$

Such a function Y is denoted as $Y(\omega) =: \mathbb{E}(X|\mathscr{G})(\omega)$. In particular, for a set $A \in \mathscr{F}$,

$$\mathbb{P}(A|\mathscr{G})(\omega) := \mathbb{E}(1_A|\mathscr{G})(\omega)$$

is called the conditional probability of A given $\mathscr{G}$.

Definition 4.1 If $p(\omega, A) := \mathbb{P}(A|\mathscr{G})(\omega)$ satisfies the following conditions, then $\mathbb{P}(A|\mathscr{G})(\omega)$ is called the *regular conditional distribution* of A given $\mathscr{G}$.

- For a.e. $\omega \in \Omega$, $A \mapsto p(\omega, A)$ is a probability measure on $(\Omega, \mathscr{F})$.
- For each $A \in \mathscr{F}$, $\omega \mapsto p(\omega, A)$ is $\mathscr{G}$-measurable.
- For any sets $A \in \mathscr{F}$ and $B \in \mathscr{G}$,

$$\mathbb{P}(A \cap B) = \int_B p(\omega, A) P(d\omega).$$

By using the regular conditional distribution, we have

$$\mathbb{E}(X|\mathscr{G})(\omega) = \int_\Omega X(\omega') p(\omega, d\omega'), \qquad a.s.$$

Hereafter, we omit (ω) to simplify the notations, provided that the omission causes no ambiguity.

Consider a real-valued stochastic process $(X_n; n \geq 1)$ in the probability space $(\Omega, \mathscr{F}, P)$.

Definition 4.2 A sequence of sub σ-fields of $\mathscr{F}$ is monotonically increasing if the following conditions hold:

- $\forall n$, $\mathscr{F}_n$ are sub σ-fields of $\mathscr{F}$;
- $\mathscr{F}_1 \subset \mathscr{F}_2 \subset \cdots$.

Such a sequence $(\mathscr{F}_n) := (\mathscr{F}_1, \mathscr{F}_2, ...)$ is called a *filtration*, or a *reference family*.

A sequence of real-valued random variables $X = (X_1, X_2, ..., X_n)$ is an $\mathbb{R}^n$-valued random variable. The smallest σ-field that makes X a random variable is then given by $\{X^{-1}(A); A \in \mathscr{B}(\mathbb{R}^n)\}$. Let us denote it by $\sigma(X_1, X_2, ..., X_n)$. In general, for a given stochastic process $(X_n; n \geq 1)$, $(\mathscr{F}_n)$ with $\mathscr{F}_n = \sigma(X_1, X_2, ..., X_n)$ is called the *natural filtration* of the process $(X_n; n \geq 1)$.

Definition 4.3 If a stochastic process $(X_n; n \geq 1)$ satisfies the following three conditions, then $(X_n; n \geq 1)$ is called a *martingale* with respect to the filtration $(\mathscr{F}_n)$, or $(\mathscr{F}_n)$-martingale.

- $(\mathscr{F}_n)$-adapted: $\forall n \in \mathbb{N}$, X_n is F_n-measurable.
- integrable: $\forall n \in \mathbb{N}$, $\mathbb{E}(|X_n|) < \infty$.
- $\forall n \in \mathbb{N}$, $\mathbb{E}(X_{n+1}|\mathscr{F}_n) = X_n$, a.s.

The first two conditions are succinctly denoted as $X_n \in L^1(\mathscr{F}_n)$.

Example 4.1 For a filtration $(\mathscr{F}_n)$ and an integrable random variable X, let $X_n = \mathbb{E}(X|\mathscr{F}_n)$. Then, (X_n) is an $(\mathscr{F}_n)$-martingale.

The next proposition is obvious.

Proposition 4.1 *If (X_n) is an $(\mathscr{F}_n)$-martingale, then*

$$\mathbb{E}(X_{n+k}|\mathscr{F}_n) = X_n, \qquad a.s., \qquad \forall n, k \in \mathbb{N}.$$

Moreover, $\mathbb{E}(X_n)$ is a constant.

A sequence of real random variables $(X_n; n \geq 1)$ is said to be *uniformly integrable* if it satisfies

$$\lim_{\lambda \to \infty} \sup_n \mathbb{E}(|X_n|, |X_n| \geq \lambda) = 0.$$

Under this condition, the tail integrals of $|X_n|$ approach zero uniformly in n. Some facts used in the following discussion are summarized in the next theorem (Theorem 2.57 of [4]), which is presented without proof.

Theorem 4.1 *Let $(X_n; n \geq 1)$ be uniformly integrable. If $X_n \to X$, a.s., then $X_n \to X$ in L^1, i.e., $\lim_{n\to\infty} \mathbb{E}(|X_n - X|) = 0$. Moreover,*

$$\lim_{n \to \infty} \mathbb{E}(X_n) = \mathbb{E}(X) < \infty.$$

Remark 4.1 Lebesgue's dominated convergence theorem posits that if $X_n \to X$, a.s. and $|X_n| \leq Y$, $\forall n \in \mathbb{N}$ for an integrable random variable Y, then $\lim_{n\to\infty} \mathbb{E}(X_n) = \mathbb{E}(X)$. The assertion is similar to Lebesgue's dominated convergence theorem, but hold under a weaker condition. In fact, $|X_n| \leq Y$ implies that $\mathbb{E}(|X_n|, |X_n| \geq \lambda) \leq \mathbb{E}(|X_n|, Y \geq \lambda) \leq \mathbb{E}(Y, Y \geq \lambda) \to 0$, as $\lambda \to \infty$.

We now present some convergence theorems of martingales. For details, see Sect. 5 of [2], Sect. 5.5 of [5], and Chap. 7 of [6]. Here, we follow the presentation of Itô [5]. Let us begin with the well-known martingale convergence theorem, which is stated without proof.

Theorem 4.2 (Martingale convergence theorem) *If a stochastic process $(X_n; n \geq 1)$ is an $(\mathscr{F}_n)$-martingale and $\sup \mathbb{E}(|X_n|) < \infty$, then as $n \to \infty$, $X_n \to X_\infty$ a.s. with $\mathbb{E}(|X_\infty|) < \infty$.*

The limit is characterized by the next theorem.

Theorem 4.3 *If a stochastic process $(X_n; n \geq 1)$ is a uniformly integrable $(\mathscr{F}_n)$-martingale, then $(X_n; n \geq 1)$ converges almost surely and in L^1. The limit $X_\infty \in L^1(\mathscr{F}_\infty)$ satisfies*

$$\mathbb{E}(X_\infty | \mathscr{F}_n) = X_n, \qquad a.s., \qquad \forall n \in \mathbb{N},$$

where $\mathscr{F}_\infty = \bigcup_n \mathscr{F}_n$.

Proof Because of the uniform integrability, $\sup \mathbb{E}(|X_n|) < \infty$. Theorem 4.2 implies almost sure convergence and $X_\infty \in L^1(\mathscr{F}_\infty)$. Moreover, Theorem 4.1 implies the L^1 convergence. For $m > n$, since $X_n = \mathbb{E}(X_m | \mathscr{F}_n)$ (Proposition 4.1), Jensen's inequality gives

$$\mathbb{E}(|X_n - \mathbb{E}(X_\infty | \mathscr{F}_n)|) = \mathbb{E}(|\mathbb{E}(X_m | \mathscr{F}_n) - \mathbb{E}(X_\infty | \mathscr{F}_n)|)$$
$$\leq \mathbb{E}(\mathbb{E}(|X_m - X_\infty| | \mathscr{F}_n)) = \mathbb{E}(|X_m - X_\infty|) \to 0,$$

as $m \to \infty$. Consequently, $X_n = \mathbb{E}(X_\infty | \mathscr{F}_n)$, $a.s.$ □

A stochastic process $(X_{-n}) := (\cdots, X_{-2}, X_{-1}, X_0)$ satisfying the conditions

$$X_{-n} \in L^1(\mathscr{F}_{-n}), \qquad \mathbb{E}(X_{-n} | \mathscr{F}_{-n-1}) = X_{-n-1}, \qquad n \in \mathbb{N}_0,$$

is called $(\mathscr{F}_{-n})$-martingale. We then have

$$\mathbb{E}(X_{-n} | \mathscr{F}_{-n-k}) = X_{-n-k}, \qquad \forall k \in \mathbb{N}, \ \forall n \in \mathbb{N}_0, \qquad a.s.$$

Theorem 4.4 (Reversed martingale convergence theorem) *If a stochastic process (X_{-n}) is an $(\mathscr{F}_{-n})$-martingale, then (X_{-n}) is uniformly integrable, and converges almost surely and in L^1. The limit $X_{-\infty} \in L^1(\mathscr{F}_{-\infty})$ satisfies*

$$X_{-\infty} = \mathbb{E}(X_{-n} | \mathscr{F}_{-\infty}), \qquad a.s., \qquad n \in \mathbb{N}_0,$$

where $\mathscr{F}_{-\infty} = \bigcap_n \mathscr{F}_{-n}$.

Proof Note that uniform integrability was assumed in Theorem 4.3. Here, we prove the uniform integrability. Since $|X_{-n}|$ is a sub-martingale with respect to $(\mathscr{F}_{-n})$,

$$\mathbb{E}(|X_{-n}|, |X_{-n}| \geq a) \leq \mathbb{E}(\mathbb{E}(|X_0| | \mathscr{F}_{-n}), |X_{-n}| \geq a) = \mathbb{E}(|X_0|, |X_{-n}| \geq a).$$

Markov's inequality gives

$$\forall n \in \mathbb{N}_0, \qquad \mathbb{P}(|X_{-n}| \geq a) \leq \mathbb{E}(|X_{-n}|)/a \leq \mathbb{E}(|X_0|)/a \to 0, \qquad a \to \infty$$

Thence, $\mathbb{E}(|X_{-n}|, |X_{-n}| \geq a) \to 0$. $\qquad\qquad\qquad\qquad\qquad\qquad\qquad$ $\square$

4.2 De Finetti's Representation Theorem

This section proves de Finetti's representation theorem (Theorem 1.1) introduced in Chap. 1. The presentation follows that of Aldous [7]. Various related topics are also covered in his book.

A regular conditional distribution defined in Definition 4.1 is a random probability measure. Let the smallest σ-field that makes α a random variable be $\sigma(\alpha)$.

Definition 4.4 For a random probability measure α, consider a sequence of random variables $(X_1, X_2, ...)$ satisfying

- $(X_1, X_2, ...)$ is conditionally independent given $\sigma(\alpha)$:

$$\mathbb{P}(X_i \in A_i; i \in [n]|\sigma(\alpha)) = \prod_{i=1}^{n} \mathbb{P}(X_i \in A_i|\sigma(\alpha)), \qquad \forall A_i \subset \mathscr{B}(\mathbb{R}), \ i \in [n], \ n \in \mathbb{N};$$

- The conditional distribution satisfies

$$\mathbb{P}(X_i \in A_i|\sigma(\alpha))(\omega) = \alpha(\omega, A_i), \qquad \forall A_i \subset \mathscr{B}(\mathbb{R}), \ i \in \mathbb{N}.$$

Such a sequence $(X_1, X_2, ...)$ is called a *mixture of i.i.d. sequences directed by α*.

According to the Glivenko–Cantelli theorem [8], the empirical distribution of an infinite i.i.d. sequence of random variables following the distribution Λ, given by

$$\Lambda_n(X_1, ..., X_n) := \frac{1}{n} \sum_{i=1}^{n} \delta_{X_i}, \qquad \delta_{X_i}(\cdot) := 1_{\{X_i \in \cdot\}}, \qquad (4.1)$$

converges uniformly to Λ; that is

$$\sup_x |\Lambda_n((-\infty, x]) - \Lambda((-\infty, x])| \to 0, \qquad a.s.,$$

as $n \to \infty$. The next lemma (Lemma 2.15 of [7]) directly follows from the Glivenko–Cantelli theorem.

Lemma 4.1 *If an infinite sequence of random variables $(X_1, X_2, ...)$ is a mixture of i.i.d. sequences, then the directing random measure is a.s. unique and is the weak limit of the empirical distribution.*

Example 4.2 Recall the Bayesian mixture modeling in Sect. 1.2.1, which models an infinite sequence of random variables $(X_1, X_2, \ldots)$ as a mixture of i.i.d. sequences

$$\mathbb{P}(X_i = x_i; i \in \mathbb{N} | \sigma(\alpha))(\omega) = \mathbb{P}(X_i = x_i; i \in \mathbb{N}; \theta_0) = \prod_{i \geq 1} \theta_0^{x_i} (1 - \theta_0)^{1 - x_i}.$$

Here, θ_0 is the realized (true) parameter of the Bernoulli trial. The strong law of large numbers implies that

$$\Lambda_n(X_1, X_2, \ldots)(\omega) \to \theta_0 \delta_1 + (1 - \theta_0) \delta_2, \qquad a.s.,$$

as $n \to \infty$. Here, $\Lambda = \theta_0 \delta_1 + (1 - \theta_0) \delta_2$ is the directing random measure. In this sense, de Finetti's representation theorem gives a condition that justifies modeling of an infinite sequence of random variables as a mixture of i.i.d. sequences.

Obviously, the random probability measure directing a mixture of i.i.d. sequences is measurable by its *tail σ-field*

$$\mathcal{T}_X := \bigcap_{n=1}^{\infty} \sigma(X_n, X_{n+1}, \ldots).$$

Moreover, the next lemma follows (Lemma 2.18 of [7]).

Lemma 4.2 *An infinite sequence of random variables $(X_1, X_2, \ldots)$ is a mixture of i.i.d. sequences if and only if $(X_1, X_2, \ldots)$ is conditionally i.i.d. given $\mathcal{T}_X$.*

Let us prepare the exchangeable σ-field.

Definition 4.5 For an infinite sequence of random variables $(X_1, X_2, \ldots)$,

$$\mathcal{E}_X := \bigcap_{n=1}^{\infty} \mathcal{E}_n, \qquad \mathcal{E}_n := \sigma(\mathcal{G}_n, X_{n+1}, X_{n+2}, \ldots),$$

with $\mathcal{G}_n(X_1, \ldots, X_n) := \sigma(f_n(X_1, \ldots, X_n));$ f_n is measurable and symmetric) is called the *exchangeable σ-field* of $(X_1, X_2, \ldots)$.

Example 4.3 A random variable is called *n*-symmetric if it is invariant under permutations of the first n variables. $\mathcal{E}_n$ is the smallest σ-field that makes all n-symmetric random variables are measurable. For example, $X_1 X_2 X_3 + X_4 X_6$ is 3-symmetric but not 4-symmetric. Therefore, it is $\mathcal{E}_3$-measurable but not $\mathcal{E}_4$-measurable.

The next proposition obviously follows from these definitions.

Proposition 4.2 $\mathcal{T}_X \subset \mathcal{E}_X$, a.s., and $\mathcal{E}_n \supset \mathcal{E}_{n+1}$, a.s.

By de Finetti's representation theorem (Theorem 1.1), an infinite exchangeable sequence is a mixture of i.i.d. sequences. The following proof comes from the proof of Theorem 3.1 in [7].

Proof (Theorem 1.1) By the exchangeability, $(X_i, Y) \overset{d}{=} (X_1, Y), i \in [n]$ for all $Y \in \mathcal{E}_n$, where $\mathcal{E}_n$ is defined in Definition 4.5. Moreover, by the symmetry, a bounded measurable function ϕ satisfies

$$\mathbb{E}(\phi(X_1)|\mathcal{E}_n) = \mathbb{E}(\phi(X_i)|\mathcal{E}_n), \quad i \in [n]$$
$$= \mathbb{E}\left(n^{-1} \sum_{i=1}^n \phi(X_i)|\mathcal{E}_n\right) = n^{-1} \sum_{i=1}^n \phi(X_i).$$

Since $\mathcal{E}_n \supset \mathcal{E}_{n+1}$, $(\mathcal{E}_1, \mathcal{E}_2, ...)$ cannot be a filtration, but $(..., \mathcal{E}_{-2}, \mathcal{E}_{-1})$ with $\mathcal{E}_{-i} := \mathcal{E}_i$ can be a filtration. Define

$$Y_{-n} := \mathbb{E}(\phi(X_1)|\mathcal{E}_{-n}).$$

Then, similarly to Example 4.1, the stochastic process $(Y_{-n}) := (\cdots, Y_{-2}, Y_{-1})$ is an $(\mathcal{E}_{-n})$-martingale. By the reversed martingale convergence theorem, we conclude that

$$n^{-1} \sum_{i=1}^n \phi(X_i) = \mathbb{E}(\phi(X_1)|\mathcal{E}_n) \equiv Y_{-n} \to \mathbb{E}(\phi(X_1)|\mathcal{E}_{-\infty}), \quad a.s., \quad (4.2)$$

as $n \to \infty$, where $\mathcal{E}_{-\infty} = \cap_{n \geq 1} \mathcal{E}_{-n}$. In the same manner, for a bounded measurable function $\phi(x_1, ..., x_k)$ with $k < n$, satisfies

$$\mathbb{E}(\phi(X_1, ..., X_k)|\mathcal{E}_n) = \frac{1}{[n]_k} \sum_{j_1=1}^n \cdots \sum_{j_k=1}^n \phi(X_{j_1}, ..., X_{j_k}) 1_{D_{n,k}}(j_1, ..., j_k),$$

where $D_{n,k}$ is a set of distinct indices. As $\#D_{n,k}^c = O(n^{k-1})$, we can conclude that

$$n^{-k} \sum_{j_1=1}^n \cdots \sum_{j_k=1}^n \phi(X_{j_1}, ..., X_{j_k}) \to \mathbb{E}(\phi(X_1, ..., X_k)|\mathcal{E}_{-\infty}), \quad a.s. \quad (4.3)$$

Suppose that $\phi(x_1, ..., x_k) = \phi_1(x_1) \cdots \phi_k(x_k)$. By using (4.2), the left-hand side of (4.3) yields

$$\prod_{i=1}^k n^{-1} \sum_{j_i=1}^n \phi_i(X_{j_i}) \to \prod_{i=1}^k \mathbb{E}(\phi_i(X_i)|\mathcal{E}_{-\infty}), \quad a.s.$$

Meanwhile, the right-hand side of (4.3) is

$$\mathbb{E}\left(\prod_{i=1}^{k}\phi_i(X_i)|\mathscr{E}_{-\infty}\right)$$

Taking $\phi_i(x_i) = 1_{\{x_i \in A_i\}}$, we conclude that $(X_1, X_2, ...)$ is a mixture of i.i.d. sequences of random variables.   □

A well-known application of de Finetti's representation theorem is *Hewitt–Savage's 0-1 law*. In Proposition 4.2, we saw that $\mathscr{T}_X \subset \mathscr{E}_X$. If $(X_1, X_2, ...)$ is exchangeable, we can show that $\mathscr{E}_X \subset \mathscr{T}_X$ and we can conclude $\mathscr{T}_X = \mathscr{E}_X$, a.s. (Corollary 3.10 of [7]). Hewitt–Savage's 0-1 law follows from this fact and the well-known Kolmogorov's 0-1 law.

Corollary 4.1 (Hewitt and Savage's 0-1 law [9]) *For an event $A \in \mathscr{E}_X$ of an i.i.d. sequence of random variables $(X_1, X_2, ...)$ is either $\mathbb{P}(A) = 0$ or $\mathbb{P}(A) = 1$.*

Example 4.4 Define a random walk $S_n := \sum_{i=1}^{n} X_i$ by an i.i.d. sequence of random variables $(X_1, X_2, ...)$. The event $\{S_n = 0, i.o.\} = \bigcap_{n=1}^{\infty}\bigcup_{m=n}^{\infty}\{S_m = 0\}$ is measurable by the tail σ-field $\mathscr{T}_S$ of $(S_1, S_2, ...)$. Here, Kolmogorov's 0-1 law is inapplicable, because $(S_1, S_2, ...)$ is not an independent sequence. Nevertheless, we cay apply the Hewitt–Savage 0-1 law to concludes that $\mathbb{P}(S_n = 0, i.o.)$ is 0 or 1, because $\{S_n = 0, i.o.\} \in \mathscr{E}_X$.

4.3 Dirichlet Process

The Dirichlet process, introduced by Ferguson in 1973, [10] is among the most fundamental prior processes in Bayesian nonparametrics. This section introduces two constructions of the Dirichlet process and discusses their relationship. Further details of the Dirichlet process, as a stochastic process, are developed in Chap. 10 of [7]. Statistical aspects are summarized in Chap. 3 of [11] and in [12]. Chapter 4 of the recent appeared book [13] provides a comprehensive introduction.

As is well known, the beta and gamma distributions are related as follows:

Proposition 4.3 *For independent random variables $X_i \sim \mathrm{Ga}(\alpha_i, 1)$, $i = 1, 2$, we have*

$$\frac{X_1}{X_1 + X_2} \sim \mathrm{Beta}(\alpha_1, \alpha_2),$$

which is independent of $X_1 + X_2 \sim \mathrm{Ga}(\alpha_1 + \alpha_2, 1)$.

The support of the m-variate Dirichlet distribution is the $(m-1)$-dimensional simplex

$$\Delta_{m-1} := \left\{ (y_1, ..., y_m); \, y_i \geq 0, \, i \in [m], \, \sum_{i=1}^{m} y_i = 1 \right\}$$ (4.4)

and its probability density function is

$$f(y_1, ..., y_m) = \frac{\Gamma(\alpha_1 + \cdots + \alpha_m)}{\Gamma(\alpha_1) \cdots \Gamma(\alpha_m)} \prod_{i=1}^{m} y_i^{\alpha_i - 1}, \qquad (y_1, ..., y_m) \in \Delta_{m-1}$$

with $\alpha_i \geq 0$, $i \in [m]$. If $\alpha_i = 0$, we regard the probability density as $\delta_0(y_i)$ by convention.

Proposition 4.4 *The Dirichlet distribution has the following properties.*

1. *For independently distributed Gamma random variables $X_i \sim \mathrm{Ga}(\alpha_i, 1)$, $i \in [m]$, we have*

$$\left(\frac{X_1}{\sum_{i=1}^{m} X_i}, \, ..., \, \frac{X_m}{\sum_{i=1}^{m} X_i} \right) \sim \mathrm{Dir}(\alpha_1, ..., \alpha_m),$$

which is independent of

$$\sum_{i=1}^{m} X_i \sim \mathrm{Ga}\left(\sum_{i=1}^{m} \alpha_i, 1 \right).$$

2. *(additivity) Dirichlet random variables $(Y_1, ..., Y_m) \sim \mathrm{Dir}(\alpha_1, ..., \alpha_m)$ satisfy*

$$\left(\sum_{i \in A_1} Y_i, \, ..., \, \sum_{i \in A_k} Y_i \right) \sim \mathrm{Dir}\left(\sum_{i \in A_1} \alpha_i, \, ... \, \sum_{i \in A_k} \alpha_i \right)$$

for any partition $\{A_1, ..., A_k\}$ of $[m]$. If we set $\alpha(A) := \sum_{i \in A} \alpha_i$ and $F(A) := \sum_{i \in A} Y_i$, then $\alpha(A)$ is a finite measure on $[m]$ and $F(A)$ is a random probability measure on $[m]$ satisfying

$$(F(A_1), ..., F(A_k)) \sim \mathrm{Dir}\left(\alpha(A_1), ..., \alpha(A_k) \right).$$

3. *For two independent m-variate Dirichlet random variables $D_\alpha \sim \mathrm{Dir}(\alpha)$ and $D_\beta \sim \mathrm{Dir}(\beta)$, and an independent beta random variable*

$$Y \sim \mathrm{Beta}\left(\sum_{i=1}^{m} \alpha_i, \sum_{i=1}^{m} \beta_i \right),$$

we have

$$Y D_\alpha + (1 - Y) D_\beta \sim \text{Dir}(\alpha + \beta).$$

4. *(conjugacy) Let $(n_1, ..., n_m)$ be a set of counts of a multinomial sampling. If the prior is $\text{Dir}(\alpha_1, ..., \alpha_m)$, the posterior is $\text{Dir}(\alpha + n)$.*

A real-valued stochastic process with stationary and independent increments has a càdlàg sample path and is called a *Lévy process*. The background is given in Chap. 2 of [14]. It is known that for any infinitely divisible distribution, there exists a Lévy process $(X_t; t \geq 0)$, $X_0 = 0$, such that X_1 follows the distribution (Theorem 7.10 of [14]). As the gamma distribution $\text{Ga}(\theta, 1)$ is infinitely divisible, there exist a Lévy process satisfying $X_1 \sim \text{Ga}(\theta, 1)$. Let us call it the *gamma (θ) process*. The characteristic function takes the following *Lévy–Khintchine representation*:

$$\mathbb{E}(e^{iuX_1}) = \exp\left\{ \int_0^\infty (e^{iux} - 1)\nu(dx) \right\}, \qquad \nu(dx) = \frac{\theta}{x}e^{-x}dx, \qquad x > 0,$$

where $i := \sqrt{-1}$. From this representation, we know that the gamma process is a pure jump process. The joint distribution of the jump size and time

$$\{(x, t); X_t - X_{t-} = x\}$$

follows the Poisson point process of intensity $\nu \times \mu$, where ν is called the *Lévy measure* and μ is the Lebesgue measure on $\mathbb{R}_{\geq 0}$ (see Theorem 19.2 of [14]). The normalized process

$$(Y_t; 0 \leq t \leq 1) = \left(\frac{X_t}{X_1}; 0 \leq t \leq 1 \right) \tag{4.5}$$

has increasing paths with interchangeable increments satisfying $Y_0 = 0$ and $Y_1 = 1$. The descending ordered countable increments $(P_1, P_2, ...)$ follow a discrete distribution whose support is the infinite-dimensional simplex

$$\Delta_\infty := \left\{ (p_1, p_2, ...); p_1 > p_2 > \cdots > 0, \sum_{i \geq 1} p_i = 1 \right\}. \tag{4.6}$$

The distribution is called the *Poisson–Dirichlet distribution* introduced in Sect. 2.3. Because of Property 1 of Proposition 4.4, we have

$$(Y_{t_1}, Y_{t_2} - Y_{t_1}, ..., Y_{t_k} - Y_{t_{k-1}}) \sim \text{Dir}(\theta t_1, \theta(t_2 - t_1), ..., \theta(1 - t_{k-1})),$$

for any sequence of times $0 < t_1 < \cdots < t_k = 1$. A sample path of the normalized process (4.5) determines a probability measure on a partition

$$\{[0, t_1], (t_1, t_2], ..., (t_{k-1}, 1]\}$$

of time $[0, 1]$. The normalized process (4.5) induces a random probability measure on $[0, 1]$, as shown in Property 2 of Proposition 4.4 for the Dirichlet distribution. More generally, for any partition

$$\{A_1 := (-\infty, z_1], A_2 := (z_1, z_2], ..., A_k := (z_{k-1}, \infty)\}$$

of $\mathbb{R}$ and a given probability measure μ on $\mathbb{R}$, a random probability measure F satisfies

$$(F(A_1), ..., F(A_k)) \sim \text{Dir}(\theta\mu(A_1), ..., \theta\mu(A_k)),$$

where $F(A_1) := Y_{\mu(z_1)}$, $F(A_k) := 1 - Y_{\mu(z_{k-1})}$, and

$$F(A_i) := Y_{\mu(z_i)} - Y_{\mu(z_{i-1})}, \quad 2 \le i \le k - 1$$

with $\mu(z) := \mu((-\infty, z])$. This is the Dirichlet process of Definition 1.3 with $\alpha = \theta\mu$. This random probability measure the *Dirichlet process* with parameter θ and the *base measure* μ. Let the Dirichlet process be denoted by $\text{DP}(\theta; \mu)$.

Remark 4.2 By its construction, the Dirichlet process has a purely atomic support. In fact, a continuous random probability measure F reduces to the base measure μ under the natural requirement that $(F(A_1), ..., F(A_k))$ is exchangeable on a partition $\{A_i; i \in [k], \mu(A_i) = 1/k\}$. This result follows because every process of the continuous-path interchangeable increments leads to an expression of the form $F([0, t]) = \alpha B_t^\circ + \beta t$, where B_t° is a Brownian bridge with $B_0^\circ = B_1^\circ = 0$. See Theorem 10.12 and the subsequent discussion in [7].

The above construction can be summarized as the following theorem by Ferguson [10].

Theorem 4.5 ([10]) *Let μ be a probability measure on a measurable space $(\mathbb{R}, \mathscr{B}(\mathbb{R}))$, and let $(X_t; t \ge 0)$ be the gamma θ process with jump sizes $(J_1, J_2, ...)$. For an i.i.d. sequence of random variables $(V_1, V_2, ...)$ following μ, we have*

$$F = \sum_{i \ge 1} \frac{J_i}{X_1} \delta_{V_i} \sim \text{DP}(\theta; \mu).$$

An alternative construction of the Dirichlet process, obtained by Sethuraman [15], is sometimes called the *stick-breaking process*. The concise proof of the following theorem is based on the description in Sect. 3.2.2 of [11].

Theorem 4.6 ([15]) *For an i.i.d. sequence of random variables* $(W_1, W_2, ...)$ *following* Beta$(1, \theta)$, *let*

$$\tilde{P}_1 = W_1, \qquad \tilde{P}_2 = (1 - W_1)W_2, \qquad ..., \qquad \tilde{P}_j = \prod_{i=1}^{j-1}(1 - W_i)W_j, \qquad ...$$

Then, for an i.i.d. sequence of random variables $(V_1, V_2, ...)$ *following* μ, *we have*

$$F = \sum_{i \geq 1} \tilde{P}_i \delta_{V_i} \sim \text{DP}(\theta; \mu).$$

Proof Given a partition $\{A_1, ..., A_k\}$, let us introduce the notation

$$\delta_{V_i}^{(k)}(A) := (\delta_{V_i}(A_1), ..., \delta_{V_i}(A_k)).$$

Letting

$$Q := \tilde{P}_1 \delta_{V_1}^{(k)} + (1 - \tilde{P}_1)U, \qquad U \sim \text{Dir}(\theta\mu(A_1), ..., \theta\mu(A_k)).$$

By Property 3 of Proposition 4.4, we obtain

$$Q|(V_1 \in A_i) \sim \text{Dir}(\theta\mu(A_1), ..., \theta\mu(A_i) + 1, ..., \theta\mu(A_k)).$$

If this expression is regarded as the posterior distribution, Property 4 of Proposition 4.4 implies that $Q \stackrel{d}{=} U$. To prove this assertion, we need to show that

$$(F(A_1), ..., F(A_k)) = \sum_{i \geq 1} \tilde{P}_i \delta_{V_i}^{(k)}(A)$$

follows the same distribution as U. As $\sum_{i=1}^{n} \tilde{P}_j + \prod_{i=1}^{n}(1 - W_j) = 1$, we have $\tilde{P}_n = W_n(1 - \sum_{i=1}^{n-1} \tilde{P}_i)$. By using this expression and the fact that $Q \stackrel{d}{=} U$, we have

$$\sum_{i=1}^{n} \tilde{P}_i \delta_{X_i}^{(k)} + \left(1 - \sum_{i=1}^{n} \tilde{P}_i\right)U = \sum_{i=1}^{n-1} \tilde{P}_i \delta_{X_i}^{(k)} + \left(1 - \sum_{i=1}^{n-1} \tilde{P}_i\right)(W_n \delta_{X_n}^{(k)} + (1 - W_n)U)$$

$$\stackrel{d}{=} \sum_{i=1}^{n-1} \tilde{P}_i \delta_{X_i}^{(k)} + \left(1 - \sum_{i=1}^{n-1} \tilde{P}_i\right)U \stackrel{d}{=} \cdots \stackrel{d}{=} U.$$

The assertion is proved in the limit $n \to \infty$ of this expression.   □

Remark 4.3 Theorem 4.6 constructs an infinite sequence of random variables, whose support is the infinite-dimensional simplex (4.6) of a sequence of independent random variables. Such a construction is called a *residual allocation model*. In particular, the law of $(\tilde{P}_1, \tilde{P}_2, ...)$ in Theorem 4.6 is called the *GEM distribution*, after Griffiths, Engen, and McCloskey.

Remark 4.4 One of the importance of Sethuraman's construction is the easiness to generate a sequence $(\tilde{P}_1, \tilde{P}_2, ...)$ on computers, because independent beta random variables are easy generated on a computer. The original construction by Ferguson requires simulating the gamma process, but generating a stochastic process is a difficult task. However, Sethuraman's construction has a difficulty in its implementation, since it is impossible to generate infinite number of random variables. Problems related to the truncated version were discussed in [16].

Comparing Theorems 4.5 and 4.6, we find that the infinite sequence of random variables $(\tilde{P}_1, \tilde{P}_2, ...)$ in Theorem 4.6 is a permutation of the sequence of random variables in the Poisson–Dirichlet distribution. Such a permutation is called the *size-biased permutation*. The size-biased permutation of a sequence $(P_1, P_2, ...)$ is generated by the following scheme:

$$\mathbb{P}(\sigma(1) = i_1) = P_{i_1},$$
$$\mathbb{P}(\sigma(j) = i_j | (\sigma(1), ..., \sigma(j-1)) = (i_1, ..., i_{j-1}))$$
$$= \frac{P_{i_j}}{1 - \sum_{k=1}^{j-1} P_{i_k}}, \qquad i_j \in \mathbb{N} \backslash \{i_1, ..., i_{j-1}\}, \qquad j = 2, 3, ...$$

The distribution of the first component, called the *structural distribution*, encodes much information about the original sequence. In Sect. 2.3.2, we discussed the asymptotics of the structural distribution of the two-parameter Poisson–Dirichlet distribution, which will be introduced in Sect. 4.5. The following proposition is a simple application of the size-biased permutation. More general result was presented in Theorem 2.2. See also Sect. 2.3 of [17].

Proposition 4.5 *For the Poisson–Dirichlet distribution with parameter θ, we have*

$$\mathbb{P}(P_1 < x) = 1 - \theta \int_x^1 y^{-1}(1 - y)^{\theta-1} dy, \qquad x > 1/2.$$

Proof Let $\tilde{\nu}_1$ denote the distribution of $\tilde{P}_1$, namely, Beta$(1, \theta)$. For a function g, we have

$$\int_0^1 g(y)\tilde{\nu}_1(dy) = \mathbb{E}(g(\tilde{P}_1)) = \mathbb{E}\left[\sum_{i \geq 1} P_i g(P_i)\right]$$

Suppose that $g(y) = 1_{\{y > x\}} y^{-1}$. Being the only component that exceed $1/2$ is P_1, the right-hand side equals to $\mathbb{E}(P_1 g(P_1)) = \mathbb{P}(P_1 > x)$ for $x > 1/2$. □

The Dirichlet process is a conjugate prior for multinomial sampling. Taking the Dirichlet process as the prior process, $F \sim \text{DP}(\theta; \mu)$, we have

$$(F(A_1), ..., F(A_k)) \sim \text{Dir}(\theta\mu(A_1), ..., \theta\mu(A_k)).$$

Owing to the conjugacy of the Dirichlet distribution for multinomial sampling (Property 4 of Proposition 4.4), the posterior is given by

$$(F_n(A_1), ..., F_n(A_k)) := (F(A_1), ..., F(A_k))|(X_1, ..., X_n)$$
$$\sim \text{Dir}\left(\theta\mu(A_1) + \sum_{i=1}^n \delta_{X_i}(A_1), ..., \theta\mu(A_k) + \sum_{i=1}^n \delta_{X_i}(A_k)\right). \qquad (4.7)$$

Therefore, the posterior is

$$F_n \sim \text{DP}\left(\theta + n, \frac{\theta\mu + \sum_{i=1}^n \delta_{X_i}}{\theta + n}\right).$$

Using this fact, let us consider the sequential sampling scheme. For simplicity, we assume that μ is diffuse (nonatomic). As

$$F(\cdot) \sim \text{Beta}(\theta\mu(\cdot), \theta(1 - \mu(\cdot))),$$

we have

$$\mathbb{P}(X_1 \in \cdot) = \mathbb{E}\{\mathbb{P}(X_1 \in \cdot|F)\} = \mathbb{E}(F(\cdot)) = \mu(\cdot).$$

Expression (4.7) then yields

$$F_1(\cdot) \sim \text{Beta}(\theta\mu(\cdot) + \delta_{X_1}(\cdot), \theta(1 - \mu(\cdot)) + (1 - \delta_{X_1}(\cdot))),$$

from which

$$\mathbb{P}(X_2 \in \cdot|X_1) = \mathbb{E}(F_1(\cdot)) = \frac{\theta\mu(\cdot) + \delta_{X_1}(\cdot)}{\theta + 1}.$$

In the same manner, we obtain

$$\mathbb{P}(X_{n+1} \in \cdot|X_1, ..., X_n) = \mathbb{E}(F_n(\cdot)) = \frac{\theta\mu(\cdot) + \sum_{i=1}^n \delta_{X_i}(\cdot)}{\theta + n}$$
$$= \frac{\theta}{\theta + n}\mu(\cdot) + \frac{n}{\theta + n}\Lambda_n(X_1, ..., X_n)(\cdot), \qquad (4.8)$$

where $\Lambda_n(X_1, ..., X_n)$ is the empirical distribution (4.1). Note that the posterior distribution is a convex combination of the prior distribution μ and the empirical distribution Λ_n, and θ represents the strength of the prior belief. The expression (4.8) is sometimes called *prediction rule*. The posterior distributions are sequentially updated by the observations. The scheme is thus called the *Bayesian updating*.

Before discussing sampling distribution from the Dirichlet process, we obtain the sampling distribution of the m-variate symmetric Dirichlet distribution. The likelihood of a multinomial sampling

$$\text{Multi}(p_1, ..., p_m), \qquad p_1 + \cdots + p_m = 1$$

is

$$\mathbb{P}(X_1 = x_1, ..., X_n = x_n; p_1, ..., p_m) = \prod_{i=1}^{n} p_i^{n_i} =: p^n,$$

where $n_i := \#\{j; X_j = i\}$. Taking the m-variate symmetric Dirichlet distribution $\text{Dir}(\alpha)$, $\alpha > 0$ as the prior distribution, the marginal likelihood is obtained as

$$\mathbb{P}_\alpha(X_1 = x_1, ..., X_n = x_n) = \frac{\Gamma(m\alpha)}{\{\Gamma(\alpha)\}^m} \int_{\Delta_{m-1}} p^{n+\alpha-1} dp = \frac{(\alpha)_{n_1} \cdots (\alpha)_{n_m}}{(m\alpha)_n}, \quad (4.9)$$

where Δ_{m-1} is the $(m-1)$-dimensional simplex defined in (4.4). The sufficient statistics are $(N_1, ..., N_m)$ and the distribution is

$$\mathbb{P}_\alpha(N_1 = n_1, ..., N_m = n_m) = \binom{-m\alpha}{n}^{-1} \prod_{i=1}^{m} \binom{-\alpha}{n_i}. \qquad (4.10)$$

The expression for $m = 2$ was previously given as (1.5). The probability mass function (4.10) is called the *Dirichlet-multinomial distribution*, or the *negative hypergeometric distribution*.

Remark 4.5 One can confirm that the m-variate symmetric Dirichlet-multinomial distribution (4.10) is a Pitman partition (2.28) with parameter $\alpha < 0$ and $\theta = -m\alpha$, $m \in \mathbb{N}$, α in (4.10) is $(-\alpha)$ in (2.28).

The sequential sampling from the Dirichlet distribution is achieved by *Pólya's urn scheme*. Consider a number of colored balls in an urn, where there are $m \in \mathbb{N}$ distinct colors. Let the number of balls of the i-th color be α_i, with $i \in [m]$. Take a ball from the urn and return the ball and another same-colored ball to the urn. After many repeats of this sampling scheme, the colors of the sampled balls $(X_1, X_2, ...)$ are distributed as

$$\mathbb{P}(X_1 = i) = \frac{\alpha_i}{\alpha_1 + \cdots + \alpha_m},$$

$$\mathbb{P}(X_{n+1} = i | X_1, ..., X_n) = \frac{\alpha_i + \sum_{j=1}^{n} \delta_{X_j}(i)}{\alpha_1 + \cdots + \alpha_m + n}. \qquad (4.11)$$

It is straightforward to confirm that the samples drawn by this sequential sampling scheme follow the probability mass function (4.9). The sequential sampling scheme for the Dirichlet process can also be represented by an urn scheme called *Blackwell–MacQueen's urn scheme* [18], also known as Hoppe's urn scheme [19] and the Chinese restaurant process [7]. Blackwell–MacQueen's urn scheme is the following algorithm.

Algorithm 4.1 ([18]) Sequential sampling from the Dirichlet process.

1. Drop a black (colorless) ball whose weight of being sampled θ into the urn.
2. Set the number of colored balls to n and the number of balls of color i to n_i. Draw one ball from the urn.

    - The black ball is drawn with probability $\dfrac{\theta}{\theta + n}$. Return the black ball to the urn, and add a ball of color taken from the continuous spectrum of colors $\mu(\cdot)$ to the urn.
    - An ball of color i is drawn with probability $\dfrac{n_i}{\theta + n}$. Return the ball to the urn, and add a ball of the same color to the urn.

3. Return to Step 2.

The sampling distribution, or the marginal likelihood of multinomial sampling from the Dirichlet process, is an analog of (4.9) for the symmetric Dirichlet distribution. If there are k distinct colors, the probability of obtaining a sample of $\{n_1, ..., n_k\}$ balls in a specific order is

$$
\frac{\theta}{\theta}\frac{1}{\theta+1}\cdots\frac{n_1-1}{\theta+n_1-1} \times \frac{\theta}{\theta+n_1}\frac{1}{\theta+n_1+1}\cdots\frac{n_2-1}{\theta+n_1+n_2-1}\cdots
$$

$$
\times \frac{\theta}{\theta+n_1+\cdots+n_{k-1}}\frac{1}{\theta+n_1+\cdots+n_{k-1}+1}\cdots\frac{n_k-1}{\theta+n-1} = \frac{\theta^k}{(\theta)_n}\prod_{i=1}^{k}(n_i-1)!.
$$

Ignoring the color labels and order of appearance, the sample $\{n_1, ..., n_k\}$ gives an integer partition of n. The probability of obtaining an integer partition λ is

$$
\mu_n(\lambda) = \frac{n!}{n_1!\cdots n_k!}\prod_{i=1}^{n}\frac{1}{c_i!} \times \frac{\theta^k}{(\theta)_n}\prod_{i=1}^{k}(n_i-1)! = \frac{n!}{(\theta)_n}\prod_{i=1}^{n}\left(\frac{\theta}{i}\right)^{c_i}\frac{1}{c_i!},
$$

where we have used the size indices $c_i(\lambda) := \#\{j; \lambda_j = i\}, i \in [n]$. This expression is the Ewens sampling formula (2.12) discussed in Chap. 2. The above derivation was given in [20].

Remark 4.6 The Ewens sampling formula can be obtained by taking the limit $m \to \infty$, $\alpha \to 0$ with $\theta \equiv m\alpha$ in (4.10). In this sense, the Dirichlet process is an infinite-dimensional version of the Dirichlet-multinomial distribution. Comparing the derivation (4.9) of the Dirichlet-multinomial distribution, we find that the Ewens sampling formula is obtained without explicitly knowing the probability density.

In Sect. 1.1, we have discussed probability measures on integer partitions. It is sometimes convenient to represent the addition rule (1.1) in terms of the probability mass function on integer partitions with size indices. In this representation, we have

$$\mu_n(c_1, ..., c_n) = \frac{c_1 + 1}{n + 1} \mu_{n+1}(c_1 + 1, ...) + \sum_{i=2}^{n+1} \frac{i(c_i + 1)}{n + 1} \mu_{n+1}(..., c_{i-1} - 1, c_i + 1, ...). \quad (4.12)$$

By a straightforward computation, one can see that the Ewens sampling formula (2.12) and the Pitman partition (2.12) have the consistency.

Let us now discuss an analog of de Finetti's representation theorem (Theorem 1.1) for infinite-exchangeable random partitions. In this analogous representation, the role played by the i.i.d. sequence of random variables is played by the *Kingman's paintbox process*. Let μ be a partly discrete and partly continuous probability measure on $[0, 1]$, and specify color-coding variable. Two colors are the same if and only if their values are the same. Imagine randomly coloring each object with an independently sampled color from μ. Let the color of the i-th object be X_i. The identity of colors $X_i = X_j$ imposes an equivalence relation $i \sim j$ between the objects. This relation induces infinitely exchangeable random partitions, and depends on the masses of the atoms of μ. Let us define a map

$$L(\mu) = p, \qquad p \in \nabla := \left\{ (p_1, p_2, ...); p_1 \geq p_2 \geq \cdots \geq 0, \sum_{i \geq 1} p_i \leq 1 \right\},$$

where p_i is the i-th largest mass, and call the infinite exchangeable random partition *paintbox p process*. The next theorem is analog of de Finetti's representation theorem, and is sometimes called Kingman's representation theorem. Aldous gave a simple proof by a trick labeling of the components by external randomization for applying de Finetti's theorem. See the proof of Proposition 11.9 in [7].

Theorem 4.7 ([21]) *Let Π be an infinite exchangeable random partition and let Π_n be its restriction to $[n]$. Also, denote by L_n a map that aligns the sizes of parts in decreasing order. The following properties then hold:*

1. *Π is a mixture of paintbox P processes.*
2. *$n^{-1} L_n(\Pi_n) \to (P_1, P_2, ...) \in \nabla$, a.s., as $n \to \infty$.*

Example 4.5 As the Ewens sampling formula is infinite exchangeable, there exists a corresponding paintbox process. Moreover, because the Ewens sampling formula is derived from the Dirichlet process, $(P_1, P_2, ...)$ in Theorem 4.7 follows the Poisson–Dirichlet distribution.

Kingman considered another natural property of random partitions called *noninterference* [21]. If a part of size r is discarded from a sample of size n, the remaining sample of size $n - r$ follows the same law for the sample of size $n - r$. In terms of size indices, the noninterference property means that

$$\frac{r c_r}{n}\mu_n(c_1, ..., c_n) = d(n, r)\mu_{n-r}(..., c_r - 1, ...),$$

where $d(n, r)$ is independent of the size indices. The following theorem (Characterization Theorem in [22]) is an important characterization of the Ewens random partition.

Theorem 4.8 ([22]) *If a random partition is infinitely exchangeable and possesses the noninterference property, it coincides with the Ewens sampling formula.*

4.4 Dirichlet Process in Bayesian Nonparametrics

Owing to its nice properties, the Dirichlet process is a fundamental prior process in Bayesian nonparametrics. Several statistical properties of the Dirichlet process are introduced through the sequential sampling scheme introduced in Sect. 4.3. We saw that the Dirichlet process is a conjugate prior for multinomial sampling. Here, we present further properties of the Dirichlet process as a prior process. Examples of use of the Dirichlet process other than in mixture models are also provided.

An infinite sequence of random variables obtained by the sequential sampling scheme is infinite exchangeable; therefore, by de Finetti's representation theorem (Theorem 1.1), it is a mixture of i.i.d. sequences. Let P_0^∞ be the product measure of the true (realized) i.i.d. sequence. The next theorem follows from Lemma 4.1.

Theorem 4.9 (Posterior consistency) *Let the prior process be $F \sim DP(\theta; \mu)$, where μ is nonatomic. Then, for diffuse P_0, we have*

$$\mathbb{P}(X_{n+1} \in \cdot | X_1, ..., X_n) = \mathbb{E}(F_n(\cdot)) \xrightarrow{d} P_0(\cdot), \qquad P_0^\infty\text{-}a.e.$$

as $n \to \infty$.

Example 4.6 ([10]) Let us consider a two-sample problem of hypothesis testing. Suppose there are two populations, one following a distribution Λ_X, the other following a distribution Λ_Y. To test the null hypothesis $\Lambda_X = \Lambda_Y$, we sample i.i.d. sequences of random variables $(X_1, ..., X_m)$ and $(Y_1, ..., Y_n)$ from Λ_X and Λ_Y, respectively. A test is constructed by estimating of the probability

$$\Delta_0 := \mathbb{P}(X_1 \le Y_1) = \int \Lambda_X d\Lambda_Y,$$

where Λ_X and Λ_Y denotes the cumulative distribution functions of Λ_X and Λ_Y, respectively. Setting the prior processes as $F_X \sim DP(\theta; \mu_X)$ and $F_Y \sim DP(\theta; \mu_Y)$ for X and Y, respectively, the prior estimate is given by

$$\int \mu_X d\mu_Y$$

and the posterior estimate is

$$\Delta_{m,n} := \int \mathbb{E}(F_{Xm})d\mathbb{E}(F_{Yn}).$$

Here, $\mathbb{E}(F_{Xm})$ and $\mathbb{E}(F_{Yn})$ are given by the prediction rules (4.8) and we have

$$\Delta_{m,n} = \frac{m}{\theta+m}\frac{n}{\theta+n}\frac{1}{mn}U + \frac{\theta}{\theta+m}\frac{\theta}{\theta+n}\int \mu_X d\mu_Y$$
$$+ \frac{\theta}{\theta+m}\frac{n}{\theta+n}\frac{1}{n}\sum_{i=1}^{n}\mu_X((-\infty, Y_i]) + \frac{m}{\theta+m}\frac{\theta}{\theta+n}\frac{1}{m}\sum_{i=1}^{m}(1-\mu_Y((-\infty, X_i])),$$

where

$$U := mn \int \Lambda_{Xm}(X_1, ..., X_m)d\Lambda_{Yn}(Y_1, ..., Y_n) = \sum_{i=1}^{m}\sum_{j=1}^{n}1_{\{X_i \leq Y_j\}}$$

is *Mann–Whitney's U statistic.* The $\Delta_{m,n}$ is a consistent estimator of Δ_0.

The posterior consistency is analogous to the law of large numbers for i.i.d. sequence of random variables. Lo [23] obtained an analog of the central limit theorem. Asymptotic normality of an estimator in the posterior distribution is described by a *Bernstein–von Mises theorem.* Chapter 12 of [13] is devoted to this subject. See Theorem 12.2 of [13] for the precise statement.

Theorem 4.10 (Bernstein–von Mises theorem) *Let the prior process be* $F \sim$ DP$(\theta; \mu)$, *where μ is diffuse. Then, for a diffuse* P_0,

$$\sqrt{n}(F_n - \mathbb{E}(F_n))(\cdot) \rightsquigarrow G_{P_0}(\cdot), \qquad P_0^\infty\text{-a.e.}$$

as $n \to \infty$. Here, G_{P_0} is the P_0-Brownian bridge with zero mean and covariances $\mathbb{E}[G_{P_0}(f)G_{P_0}(g)] = P_0(fg) - P_0(f)P_0(g)$, *where* $P_0(f) = \int f(x)P_0(dx)$.

Example 4.7 According to Theorem 4.9, $\mathbb{E}(F_n(\cdot))$ provides a consistent point estimate of the true probability measure $P_0(\cdot)$. Let us consider an approximate credible set of the estimation. Theorem 4.10 and properties of the Brownian bridge, we obtain the asymptotic $(1-\alpha)$-simultaneous credible set

$$\left[\mathbb{E}(F_n((-\infty, x])) - \frac{\lambda}{\sqrt{n}}, \mathbb{E}(F_n((-\infty, x])) + \frac{\lambda}{\sqrt{n}}\right], \qquad \forall x \in \mathbb{R},$$

where λ solves the expression

$$2\sum_{i\geq 1}(-1)^{i+1}e^{-2i^2\lambda^2} = \alpha.$$

If θ tends to zero, this Bayesian credible set reduces to the simultaneous confidence set given by the Kolmogorov–Smirnov test.

Estimating the number of unseen species is a classical subject in Bayesian statistics. One of the early studies is done by Efron and Thisted on the estimation of the number of words did Shakespeare knows [24]. Recent progress on the subject can be found in [25–27]) and reference therein.

Example 4.8 Keener et al. studied this problem with an empirical Bayes approach based on the Dirichlet-multinomial model and Dirichlet process [28]. Suppose frequencies of species in a population follow the m-variate symmetric Dirichlet distribution of parameter $\alpha > 0$. Then, the marginal likelihood is the Dirichlet-multinomial distribution (4.10). Note that the total number of species m cannot be observed; the number of observed species is the number of nonzero n_i in (4.10). It is the sufficient statistic of m. If the parameter α is known, the UMVUE of m can be constructed by using the result in Proposition 3.4.

4.5 Related Prior Processes

Let us see some prior processes beyond the Dirichlet process and revisit some measures on partitions discussed in Chap. 2. As mentioned in Sect. 4.3, prior processes can be modeled with prediction rules or with normalized Lévy processes. Models beyond the Dirichlet process are concisely surveyed in [29]. This monograph we will concentrate on random probability measures associated with normalized subordinators (increasing Lévy processes). However, various random measures are also used as prior processes in Bayesian nonparametrics. For examples, see [30, 31].

Theorem 4.8 concludes by assuming that both of infinite exchangeability and non-interference property uniquely identifies the prior process as a Dirichlet process. In usual statistical inferences, sizes of samples are arbitrary. Hence, infinite exchangeability is desirable. This chapter considers models with infinite exchangeability.

4.5.1 Prediction Rules

The present subsection discusses prediction rules. Prior processes given by prediction rules are called *species sampling priors*. The prediction rule exists for any given exchangeable partition probability function (EPPF). Let us consider a sequential sampling scheme from a prior process F. Let the prior distribution μ be diffuse and given by

$$\mathbb{P}(X_1 \in \cdot) = \mathbb{E}(F(\cdot)) = \mu(\cdot).$$

The posterior distribution, or the prediction rule, for a given sample $(X_1, ..., X_n)$ is given by

$$\mathbb{P}(X_{n+1} \in \cdot | X_1, ..., X_n) = \left(1 - \sum_{i=1}^{K_n} f_i(n_1, ..., n_{K_n})\right) \mu(\cdot) + \sum_{i=1}^{K_n} f_i(n_1, ..., n_{K_n}) \delta_{Z_i}(\cdot), \quad (4.13)$$

where Z_i is the i-th distinct value in $(X_1, ..., X_n)$ and K_n is the number of distinct values in $(X_1, ..., X_n)$. If the sequence of random variables is exchangeable, we can apply the consistency condition (1.1) for EPPF p_n to get

$$1 - \sum_{i=1}^{k} f_i(n_1, ..., n_k) = \frac{p_{n+1}(n_1, ..., n_k, 1)}{p_n(n_1, ..., n_k)}, \quad f_i(n_1, ..., n_k) = \frac{p_{n+1}(n_1, ..., n_i + 1, ..., n_k)}{p_n(n_1, ..., n_k)}.$$

Remark 4.7 In Bayesian mixture models, prediction rules have been used as the updating rules of Gibbs samplers. An earlier application of the Dirichlet process is described in Escobar and West [32]. In mixture models, a sample is not $(X_1, ..., X_n)$ (see Remark 1.1) and we need a prediction rule of Y_{n+1} given sample $(Y_1, ..., Y_n)$. The prediction rule involves integration of $\mathbb{P}(Y|X)$ with $\mathbb{P}(X)$ and this modification causes difficulties, such as non-conjugacy and slow mixing. These issues for general species sampling priors are extensively discussed in [16, 33].

An EPPF gives a prediction rule, but the reverse is not always true. Lee et al. [34] established the following theorem.

Theorem 4.11 ([34]) *The necessary and sufficient conditions under which a prediction rule (4.13) gives an EPPF is*

$$f_i(n_1, ..., n_k) f_j(n_1, ..., n_i + 1, ..., n_k) = f_j(n_1, ..., n_k) f_i(n_1, ..., n_j + 1, ..., n_k)$$

and

$$f_i(n_1, ..., n_k) = f_{\sigma^{-1}(i)}(n_{\sigma(1)}, ..., n_{\sigma(k)})$$

for all permutations σ of $[k]$. Here, $(n_1, ..., n_k)$, i, and j are arbitrary.

The Gibbs partitions (2.21) introduced in Chap. 2 are fundamental models of EPPFs. The probability mass function is

$$\mu_n(\lambda) = \frac{v_{n,l(\lambda)} n!}{B_n(v, w)} \prod_{i=1}^{n} \left(\frac{w_i}{i!}\right)^{c_i} \frac{1}{c_i!}, \quad \lambda \in \mathscr{P}_n, \quad (4.14)$$

or equivalently,

$$p_n(n_1, ..., n_{l(\lambda)}) = \frac{v_{n,l(\lambda)}}{B_n(v, w)} \prod_{i=1}^{l(\lambda)} w_{n_i}. \quad (4.15)$$

A Gibbs partition is not always infinitely exchangeable.

Example 4.9 (*Pitman partition*) The Pitman partition, which was extensively discussed in Chap. 2, is a well-known infinite exchangeable Gibbs partition. The species sampling prior characterized by the Pitman partition is called the *two-parameter Poisson–Dirichlet process*, or the *Pitman–Yor process* [16, 35]. Let it be denoted by $DP(\alpha, \theta; \mu)$. The prediction rule is given by

$$1 - \sum_{i=1}^{k} f_i(n_1, ..., n_k) = \frac{\theta + k\alpha}{\theta + n}, \qquad f_i(n_1, ..., n_k) = \frac{n_i - \alpha}{\theta + n}. \qquad (4.16)$$

It is straightforward to confirm that this prediction rule gives the Pitman partitions, as obtained for the Ewens sampling formula in Sect. 4.3.

After constructing a two-parameter Poisson–Dirichlet process with size-biased permutation, we can establish the following theorem (Theorem 3.8 in [36]), which implies that the two-parameter Poisson–Dirichlet process is conjugate to multinomial sampling if and only if $\alpha = 0$.

Theorem 4.12 ([36]) *Let the prior process be* $F \sim DP(\alpha, \theta; \mu)$, *where* μ *is diffuse. The posterior is given by*

$$F|(X_1, ..., X_n) \sim U_{K_n+1}DP(\alpha, \theta + K_n\alpha; \mu) + \sum_{i=1}^{K_n} U_i\delta_{Z_i},$$

where Z_i *is the* i-*th distinct value in* $(X_1, ..., X_n)$, K_n *is the number of distinct values in* $(X_1, ..., X_n)$, *and*

$$(U_1, ..., U_{K_n}, U_{K_n+1}) \sim \text{Dir}(n_1 - \alpha, ..., n_{K_n} - \alpha, \theta + K_n\alpha).$$

Remark 4.8 Similar to the prediction rule of the Dirichlet process in Sect. 4.3. the prediction rule of the two-parameter Poisson–Dirichlet process (4.16) is immediately obtained from the posterior.

By similar arguments to those of Theorem 4.9, we obtain the following theorem.

Theorem 4.13 ([37, 38]) *Let the prior process be* $F \sim DP(\alpha, \theta; \mu)$, *where* μ *is diffuse. Then, for some diffuse* P_0, F *has posterior consistency if and only if* $\alpha = 0$.

Remark 4.9 A necessary and sufficient condition for a species sampling prior with posterior consistency is $\sum_{i=1}^{k} f_i(n_1, ..., n_k) \to 0$ as $n \to \infty$. See Theorem 4 of [38].

Remark 4.10 Note that consistency depends on the context. Let us consider a Bayesian mixture model: $Y_i|(X_i, \sigma^2) \sim N(\mu_{X_i}, \sigma^2)$, where $(X_1, ..., X_n)$ follows (4.13). Lo [39] discussed this model with the Dirichlet process. Consistency is achieved if the posterior probability of the mixture distribution with a prior process F:

$$\int_U \phi_\sigma(y - x) dF(x)$$

given observations $(Y_1, ..., Y_n)$ converges to 1 for all neighborhoods U of the true probability measure P_0 as $n \to \infty$, where $\phi_\sigma(y - x)$ is the normal density with mean 0 and variance σ^2. This consistency has been established for the Dirichlet process by Ghosal et al. [40], but it holds for more general classes of prior processes, including the two-parameter Poisson–Dirichlet process [41].

Gnedin and Pitman [42] gave the following characterization of Gibbs partitions with infinite exchangeability. The second assertion was given by Kerov [43].

Lemma 4.3 ([42]) *A Gibbs partition (4.14) is infinite exchangeable if and only if the sequence (w_i) satisfies*

$$w_i = (\beta - \alpha)_{i-1;\beta}, \quad i \in \mathbb{N}, \quad \alpha < \beta, \quad \beta \geq 0.$$

In particular, if the sequence $(v_{n,k})$ does not depend on n, then a Gibbs partition with infinite exchangeability is the Pitman partition (2.28).

Proof As the probability mass function is invariant under a change of sequences $(w_i, v_{n,k}) \mapsto (tw_i, t^{-k}v_{n,k})$, we may set $w_1 = 1$. The consistency condition (4.12) for infinite exchangeability is

$$\frac{v_{n,k}}{B_n(v, w)} = \frac{v_{n+1,k}}{B_{n+1}(v, w)} \sum_{i=1}^n r_i c_i + \frac{v_{n+1,k+1}}{B_{n+1}(v, w)}, \quad r_i \equiv \frac{w_{i+1}}{w_i}, \quad k \in [n].$$

Now set $k = 2$. As $r_i + r_{n-i}$ regardless of i, r_i is an arithmetic series. Putting $r_i = \beta i - \alpha, \alpha < \beta, \beta \geq 0$, we obtain $w_i = (\beta - \alpha)_{i-1;\beta}$. As $\sum_{i=1}^n r_i c_i = \beta n - \alpha k$, we have for the latter assertion

$$\frac{v_{k+1}}{v_k} - \alpha k = \frac{B_{n+1}(v, w)}{B_n(v, w)} - \beta n = const. \equiv \theta,$$

where v_k denotes $v_{n,k}$. Moreover, as the probability mass function is invariant under the change of sequence $v_k \mapsto tv_k$, we may set $v_1 = \theta$. We, then, obtain $v_k = (\theta)_{k;\alpha}$ and $B_n(v, w) = (\theta)_{n;\beta}$. After normalization, we obtain $\beta = 1$, which completes the proof. $\square$

Remark 4.11 By Lemma 4.3, Kolchin's model introduced in Sect. 2.2 reduces to the Pitman partition under the infinite exchangeability.

Moreover, under the change of sequence $(w_{n,k})$ as $v_{n,k} \mapsto v_{n,k}\beta^{-k}$, a Gibbs partition with infinite exchangeability in Lemma 4.3 can be represented by a sequence (w_i) of the form

$$w_i = (1 - \alpha)_{i-1}, \qquad i \in \mathbb{N}, \qquad \alpha < 1. \tag{4.17}$$

In the following discussion, we assume (4.17) for Gibbs partitions with infinite exchangeability. To simplify the expressions, we also set

$$\tilde{v}_{n,k} \equiv \frac{v_{n,k}}{B_n(v, w)}.$$

Gnedin and Pitman characterized the exchangeable Gibbs partitions by the following theorem (Theorem 12 in [42]).

Theorem 4.14 ([42]) *An infinite exchangeable Gibbs partition of the form (4.14) with a sequence (w_i) of the form (4.17) is one or a mixture of the following partitions:*

- *The Ewens sampling formula (2.12), if $\alpha = 0$ and $\theta > 0$.*
- *The m-variate symmetric Dirichlet-multinomial distribution (4.10), if $\alpha < 0$ and $\theta = -m\alpha$, $m \in \mathbb{N}$.*
- *A Poisson–Kingman partition associated with the α-stable subordinator in Definition 4.9 (in the next subsection), if $\alpha \in (0, 1)$.*

Remark 4.12 Some of the Gibbs partitions, such as the limiting quasi-multinomial distribution (Example 2.14), the limiting conditional inverse Gaussian–Poisson distribution (Remark 2.7), and the multiplicative measure induced by the exponential structure associated with the Macdonald symmetric functions (Example 2.15), are not infinite exchangeable. Moreover, when the model involves dependence structures, such as regression, hierarchy, and spatiotemporal correlation, exchangeability cannot be justified. Such modelings are discussed in [44–48].

As mentioned in Sect. 4.3, the prediction rule (4.8) of the Dirichlet process gave us a sampling scheme known as Blackwell–MacQueen's urn scheme (Algorithm 4.1). For an infinite-exchangeable Gibbs partition, the prediction rule (4.13) gives the following modified Blackwell–MacQueen's urn scheme. Note that this algorithm demands infinite exchangeability.

Algorithm 4.2 ([49]) Sequential sampling from an infinite-exchangeable Gibbs partition.

1. Drop a black (colorless) ball into the urn.
2. Set the number of colored balls to n and the number of balls of color i to n_i. Draw one ball from the urn.

    - The black ball is drawn with probability $\dfrac{\tilde{v}_{n+1,k+1}}{\tilde{v}_{n,k}}$. Return the black ball to the urn, and add a ball with a color sampled from the continuous spectrum of colors $\mu(\cdot)$ to the urn.

- A ball of color i is drawn with probability $\dfrac{\tilde{v}_{n+1,k}}{\tilde{v}_{n,k}}(n_i - \alpha)$. Return the colored ball to the urn, and add a ball of the same color to the urn.

3. Return to Step 2.

Example 4.10 (*Pitman partitions*) The prediction rule (4.16) of the Pitman partition gives the probabilities in Step 2.

Example 4.11 (*Generalized gamma NRMI*) Next subsection introduces the generalized gamma NRMI, the only NRMI whose EPPF is an infinite exchangeable Gibbs partition. The expressions for $\tilde{v}_{n,k}$ for the generalized gamma NRMI, which involve sums of incomplete gamma functions, are expressed in (4.25). Numerical evaluations are discussed in [50].

Example 4.12 (*Poisson–Kingman partitions*) For the conditional Poisson–Kingman partition associated with the stable subordinator of parameter $1/2$ (see next subsection), the probabilities in Step 2 of Algorithm 4.2 are given by

$$\frac{z h_{k-2n}(z)}{h_{k-2n+1}(z)}, \qquad \frac{2 h_{k-2n-1}(z)}{h_{k-2n+1}(z)}\left(n_i - \frac{1}{2}\right),$$

where $h_i(z)$ satisfies the recurrence relation

$$h_{i+1}(z) = z h_i(z) - i h_{i-1}(z), \qquad h_0(z) = 1,$$
$$h_{-1}(z) = e^{z^2/2}\int_z^\infty e^{-x^2/2}dx.$$

For $i \in \mathbb{N}$, $h_i(z)$ are the Hermite polynomials, which are orthogonal with respect to the standard normal density.

4.5.2 Normalized Subordinators

This subsection discusses the modeling of prior processes by normalizing a subordinator. Especially, it introduces the normalized random measure with independent increments (NRMI) and the Poisson–Kingman partition.

In Sect. 4.5.1, we introduced the two-parameter Poisson–Dirichlet process, which can be constructed by normalizing the stable subordinator. See Chap. 3 of [51] for the detailed explanation. Let Z, Z_1, ..., Z_n be independent random variables following the α-*stable distribution*, whose Laplace transform satisfies $\mathbb{E}(\exp(-\lambda Z)) = \exp(-\lambda^\alpha)$. We then have

$$n^{-1/\alpha}(Z_1 + Z_2 + \cdots + Z_n) \overset{d}{=} Z, \qquad \alpha \in (0, 2].$$

The stable distributions of $\alpha = 2$ and $\alpha = 1$ are the normal distribution and the Cauchy distribution, respectively. We here consider the case of $\alpha \in (0, 1)$. The probability density function of the α-stable distribution f_α was given in (2.47). The distribution of $Z^{-\alpha}$ is the Mittag-Leffler distribution with the p-th moment of $\Gamma(1 + p)/\Gamma(1 + p\alpha)$. The α-stable subordinator is a process $(X_t; t \geq 0)$ with $X_0 = 0$ and X_1 follows the α-stable distribution with the Lévy measure of

$$\nu(dx) = \frac{\alpha x^{-(1+\alpha)}}{\Gamma(1 - \alpha)} dx, \qquad x > 0. \tag{4.18}$$

The normalized process $(Y_t; 0 \leq t \leq 1) := (X_t/X_1; 0 \leq t \leq 1)$ has interchangeable positive increments satisfying $Y_0 = 0$ and $Y_1 = 1$. The descending ordered countable increments $(P_1, P_2, ...)$ follow a discrete distribution whose support is the infinite-dimensional simplex in (4.6). This distribution is called the *two-parameter Poisson–Dirichlet distribution* of parameters α and $\theta = 0$. For $\alpha \in (0, 1)$ and $\theta > -\alpha$, the two-parameter Poisson–Dirichlet distribution $\mathbb{P}_{\alpha,\theta}$ is obtained by the measure changing formula:

$$\frac{d\mathbb{P}_{\alpha,\theta}}{d\mathbb{P}_{\alpha,0}} = \frac{\Gamma(\theta + 1)}{\Gamma(\theta/\alpha + 1)} X_1^{-\theta}. \tag{4.19}$$

Remark 4.13 The two-parameter Poisson–Dirichlet distributions give the distributions of sequences of ranked excursion lengths of a standard Bessel process [17, 35, 52]. For a standard Bessel process of dimension $2 - 2\alpha$, up to time 1 and including the meander length, the sequence follows the two-parameter Poisson–Dirichlet distribution of parameters α and $\theta = 0$. For a standard Bessel bridge of dimension $2 - 2\alpha$, the sequence follows the two-parameter Poisson–Dirichlet distribution with parameters α and $\theta = \alpha$.

The two-parameter Poisson–Dirichlet process introduced in Sect. 4.5.1 is defined in terms of the two-parameter Poisson–Dirichlet distribution as follows:

Definition 4.6 ([35]) Let μ be a probability measure on a measurable space $(\mathbb{R}, \mathscr{B})$ and let $(P_1, P_2, ...)$ be a sequence of random variables following the two-parameter Poisson–Dirichlet distribution with parameters $\alpha \in (0, 1)$ and $\theta > -\alpha$. For an i.i.d. sequence of random variables $(V_1, V_2, ...)$ following μ, we have

$$F = \sum_{i \geq 1} P_i \delta_{V_i} \sim DP(\alpha, \theta; \mu).$$

The stick-breaking construction (see Theorem 4.6) is defined as follows.

Theorem 4.15 *For a sequence of independent random variables* $(W_1, W_2, ...)$ *with* $W_i \sim Beta(1 - \alpha, \theta + i\alpha)$, *let*

$$\tilde{P}_1 = W_1, \qquad \tilde{P}_2 = (1 - W_1)W_2, \qquad \cdots \qquad \tilde{P}_j = \prod_{i=1}^{j-1}(1 - W_i)W_j, \ldots$$

Then, for an i.i.d. sequence of random variables $(V_1, V_2, \ldots)$ following μ, we have

$$F = \sum_{i \geq 1} \tilde{P}_i \delta_{V_i} \sim DP(\alpha, \theta; \mu).$$

Here, the sequence $(\tilde{P}_1, \tilde{P}_2, \ldots)$ is a size-biased permutation of a sequence following the two-parameter Poisson–Dirichlet distribution $(P_1, P_2, \ldots)$. The following sketch of a proof derives from Sects. 4.1 and 4.2 of [17, 52, 53].

Proof Consider a subordinator $(X_t; t \geq 0)$ with $X_0 = 0$ with the Lévy density of $v(dx) = \rho(x)dx$. Let the probability density function of X_1 be $f_{X_1}(x)$. Then, for $\lambda \geq 0$ the Lévy–Khintchine representation is

$$\mathbb{E}(e^{-\lambda X_1}) = \int_0^\infty e^{-\lambda x} f_{X_1}(x)dx = \exp(-\Psi(\lambda)), \qquad (4.20)$$

where the Laplace exponent is

$$\Psi(\lambda) := \int_0^\infty (1 - e^{-\lambda x})\rho(x)dx. \qquad (4.21)$$

Here, f_{X_1} is uniquely determined by solving the following integral equation:

$$f_{X_1}(x) = \int_0^x \rho(s) f_{X_1}(x - s)\frac{s}{x}ds \qquad (4.22)$$

The Laplace transform of this integral equation is obtained by differentiating (4.20) with respect to λ. Let the descending ordered jump sizes be denoted by $(J_1, J_2, \ldots)$ and let the size-biased permutation of $(J_1/X_1, J_2/X_1, \ldots)$ be denoted by $(\tilde{J}_1/X_1, \tilde{J}_2/X_1, \ldots)$. The right-hand side of the expression (4.22) has a simple interpretation. The jump sizes follow the Poisson point process of intensity v. Now, decompose x into a randomly picked jump size s and the remainder $x - s$, where the jump is picked by size-biased sampling. The probability that a jump size s is taken is s/x, whose density is $\rho(s)$. Therefore, the joint density of $\tilde{J}_1$ and X_1 is given as

$$f_{\tilde{J}_1, X_1}(s, x) = \rho(s) f_{X_1}(x - s)\frac{s}{x},$$

The joint density of $\tilde{X}_1 \equiv X_1 - \tilde{J}_1$ and $U_1 \equiv 1 - \tilde{J}_1/X_1$ is

$$f_{\tilde{X}_1, U_1}(x, u) = u^{-1}\rho^*((1 - u)u^{-1}x) f_{X_1}(x),$$

where we set $\rho^*(x) \equiv x\rho(x)$. Especially, for the α-stable subordinator, we have

$$f_{\tilde{X}_1, U_1}(x, u) \propto (1 - u)^{-\alpha} u^{\alpha-1} x^{-\alpha} f_\alpha(x).$$

This implies that $W_1 \equiv 1 - U_1$ and $\tilde{X}_1$ are independent, $W_1 \sim Beta(1 - \alpha, \alpha)$, and $f_{\tilde{X}_1}(x) \propto x^{-\alpha} f_\alpha(x)$. In the same manner, we obtain

$$f_{\tilde{J}_1, \tilde{J}_2, X_1}(s, t, x) = \rho(s)\rho(t) f_{X_1}(x - s - t) \frac{s}{x} \frac{t}{x - s}$$

and for the α-stable subordinator, we have

$$f_{\tilde{X}_2, U_1, U_2}(x, u, v) \propto (1 - u)^{-\alpha} u^{\alpha-1} (1 - v)^{-\alpha} v^{2\alpha-1} x^{-2\alpha} f_\alpha(x),$$

where $U_2 \equiv 1 - \tilde{J}_2/\tilde{X}_1$ and $\tilde{X}_2 \equiv \tilde{X}_1 - \tilde{J}_2$. This expression implies the independence of W_1, W_2, and $\tilde{X}_2$. Here, $W_2 \equiv 1 - U_2 \sim Beta(1 - \alpha, 2\alpha)$, and $f_{\tilde{X}_2} \propto x^{-2\alpha} f_\alpha(x)$. This procedure gives the assertion of the theorem for $\theta = 0$. The assertion for $\theta \neq 0$ follows immediately by applying the measure changing formula (4.19) to the density f_{X_1}. ☐

Remark 4.14 In Remark 4.3, the sequence obtained by size-biased permutation of a sequence of random variables following the Poisson–Dirichlet distribution was called the GEM distribution. The size-biased permutation of a sequence following the two-parameter Poisson–Dirichlet distribution follows the two-parameter GEM distribution, the only residual allocation model with invariance under biased permutations [54].

The NRMI is defined as follows.

Definition 4.7 Let μ be a finite measure on a measurable space $(\mathbb{R}, \mathscr{B}(\mathbb{R}))$ such that for any $A_1, ..., A_n \in \mathbb{R}$, $A_i \cap A_j = \phi, \forall i \neq j$, the random variables $\mu(A_1), ..., \mu(A_n)$ are mutually independent. Then, μ is called a complete random measure. The random probability measure $\mu/\mu(\mathbb{R})$ is called a normalized random measure with independent increments.

Let us consider complete random measures $\mu = \sum_{i \geq 1} J_i \delta_{V_i}$ on $\mathbb{R}$, where the positive jumps J_i and the $\mathbb{R}$-valued locations V_i are both random. If the jump-size distribution is independent of their locations, the random measure μ is called homogeneous. Pitman [53] showed that a homogeneous NRMI is a species sampling prior with the following EPPF:

$$p(n_1, ..., n_k) = \frac{(-1)^{n-k}}{(n-1)!} \int_0^\infty \lambda^{n-1} e^{-\Psi(\lambda)} \prod_{i=1}^k \Psi^{(n_i)}(\lambda) d\lambda, \qquad (4.23)$$

where $\Psi^{(i)}(\lambda)$ is the i-th derivative of the Laplace exponent (4.21). Lijoi et al. proved that an NRMI whose EPPF is an infinite exchangeable Gibbs partition reduces to the following generalized gamma NRMI [55].

Definition 4.8 ([55]) An NRMI with Lévy density

$$\nu(dx) = \frac{x^{-(1+\alpha)}\theta e^{-\tau x}}{\Gamma(1-\alpha)} dx, \qquad x > 0, \qquad \alpha \in (0,1), \ \tau > 0, \ \theta > 0 \quad (4.24)$$

is called a generalized gamma NRMI.

A straightforward observation confirms that the EPPF of the generalized Gamma NRMI is an infinite exchangeable Gibbs partition. Substituting the Lévy measure (4.24) into (4.23), we have

$$p(n_1, ..., n_k) = \frac{e^\beta \alpha^{k-1}}{(n-1)!} \sum_{i=0}^{n-1} \binom{n-1}{i} (-1)^i \beta^{i/\alpha} \Gamma\left(k - \frac{i}{\alpha}, \beta\right) \prod_{j=1}^{k} (1-\alpha)_{n_j-1},$$

where

$$\Gamma(z, \beta) = \int_\beta^\infty e^{-t} t^{z-1} dt, \qquad \beta \equiv \frac{\theta \tau^\alpha}{\alpha}.$$

Comparing this expression with (4.15), we obtain (4.17) and the following expression:

$$\tilde{\nu}_{n,k} = \frac{e^\beta \alpha^{k-1}}{(n-1)!} \sum_{i=0}^{n-1} \binom{n-1}{i} (-1)^i \beta^{i/\alpha} \Gamma\left(k - \frac{i}{\alpha}, \beta\right). \qquad (4.25)$$

Example 4.13 Taking $\theta = \alpha$ and $\tau = 0$ in (4.24) yields the two-parameter Poisson–Dirichlet process with parameters $\alpha \in (0,1)$ and $\theta = 0$. Taking $\alpha = 0$ and $\tau = 1$ yields the Dirichlet process with parameter θ.

Example 4.14 The probability distribution function of the stable distribution with parameter $\alpha = 1/2$ (2.47) takes the following closed form:

$$f_{1/2}(x) = \frac{x^{-3/2}}{2\sqrt{\pi}} e^{-\frac{1}{4x}} = \mathbb{P}\left(B_1^{-2}/2 \in (x, x + dx]\right)/dx, \qquad B_1 \sim N(0,1),$$

The generalized Gamma NRMI with $\alpha = \tau = 1/2$ is called the *normalized-inverse Gaussian process*, because $X_1 \sim IG(\sqrt{2\theta}, 1)$, where the probability density function of the inverse Gaussian distribution $IG(\beta, 1)$ is

$$f(x) = \frac{\beta}{\sqrt{2\pi}} x^{-\frac{3}{2}} \exp\left\{-\frac{1}{2}\left(\frac{\beta^2}{x} + x\right) + \beta\right\}, \qquad x > 0.$$

The properties of the normalized-inverse Gaussian process and applications to density estimation are extremely discussed in [56].

The Poisson–Kingman partition appeared in Theorem 4.14 characterized the exchangeable Gibbs partition. The definition is as follows:

Definition 4.9 Consider a homogeneous NRMI given by a subordinator with Lévy measure v. The random measure obtained by conditioning with $X_1 = x$ is denoted by $\mathbb{P}_{v|x}$. The random measure obtained as the mixture of $\mathbb{P}_{v|x}$ with the mixing distribution $\eta(dx)$:

$$\int_0^\infty \mathbb{P}_{v|x}(\cdot)\eta(dx)$$

is called the *Poisson–Kingman distribution* with parameter v and η. The EPPF of the Poisson–Kingman distribution is called the *Poisson–Kingman partition*.

Example 4.15 If $X_1 \sim \eta(x)$, the Poisson–Kingman distribution is a homogeneous NRMI. If the EPPF gives the Gibbs partition, it reduces to a generalized gamma NRMI. In Example 4.13, we showed that the two-parameter Dirichlet process with parameters $\alpha \in (0, 1)$ and $\theta = 0$ is a generalized gamma NRMI, given as a mixture of $\mathbb{P}_{v|x}$ whose v is the α-stable distribution (4.18) and the mixing distribution η are α-stable distributions.

Example 4.16 The two-parameter Poisson–Dirichlet distribution with parameters $\alpha \in (0, 1)$ and $\theta > -\alpha$ is constructed as the Poisson–Kingman distribution. A mixture of $\mathbb{P}_{v|x}$ whose v is the α-stable distribution (4.18) and whose mixing distribution is given by

$$\eta(dx) = \frac{\Gamma(1+\theta)}{\Gamma(1+\theta/\alpha)} x^{-\theta} f_\alpha(x)dx.$$

This expression follows from the measure changing formula (4.19). Comparing this result with (2.1) we find that $X_1 \overset{d}{=} S_\alpha^{-1/\alpha}$, where S_α is the scaled limit of the length of the Pitman partition.

As the Poisson–Kingman distribution is a mixture of $\mathbb{P}_{v|x}$ whose v is the α-stable distribution, we can represent the Poisson–Kingman partition as a mixture of conditional partitions given $X_1 = x$, where X_1 follows an α-stable distribution. The conditional partition is also an infinite exchangeable EPPF with the following form [53]:

$$\tilde{v}_{n,k}|x = \frac{\alpha^k x^{-n}}{\Gamma(n-k\alpha)f_\alpha(x)}\left\{\int_0^x y^{n-1-k\alpha} f_\alpha(x-y)dy\right\}.$$

For $\alpha = 1/2$, the expression reduces to

$$2^{n-k}z^{k-1}h_{k+1-2n}(z), \qquad z \equiv 1/\sqrt{2x},$$

where $h_i(z)$ are the Hermite polynomials defined in Example 4.12. Moreover, Ho et al. [57] derived closed expressions for general $\alpha \in (0, 1)$ in terms of some special functions.

References

1. Billingsley, P.: Probability and Measure, 3rd edn. Wiley, New York (1995)
2. Durrett, R.: Probability: Theory and Examples, 4th edn. Cambridge University Press, New York (2010)
3. Rogers, L.C.G., Williams, D.: Diffusions, Markov Processes and Martingales, vol. 1, 2nd edn. Cambridge University Press, New York (2000)
4. Funaki, T.: Kakuritsu-ron. Asakura, Tokyo (2004). in Japanese
5. Itô, K.: Kakuritsu-ron. Iwanami, Tokyo (1991). in Japanese
6. Doob, J.L.: Stochastic Processes. Wiley, New York (1953)
7. Aldous, D.J.: Exchangeability and related topics. In: Ecole d'Été de Probabilités de Saint Flour. Lecture Notes in Mathematics, vol. 1117. Springer, Berlin (1985)
8. Lehmann, E.L., Romano, J.P.: Testing Statistical Hypotheses, 3rd edn. Springer, New York (2005)
9. Hewitt, E., Savage, L.J.: Symmetric measures on Cartesian products. Trans. Am. Math. Soc. **80**, 470–501 (1955)
10. Ferguson, T.S.: A Bayesian analysis of some nonparametric problems. Ann. Stat. **1**, 209–230 (1973)
11. Ghosh, J.K., Ramamoorthi, R.V.: Bayesian Nonparametrics. Springer, New York (2003)
12. Ghosal, S.: The Dirichlet process, related priors and posterior asymptotics. In: Hjort, N.L., Holmes, C., Müller, P., Walker, S.G. (eds.) Bayesian Nonparametrics. Cambridge University Press, Cambridge (2010)
13. Ghosal, S., van der Vaart, A.: Fundamentals of Nonparametric Bayesian Inference. Cambridge University Press, Cambridge (2017)
14. Sato, K.-I.: Lévy Processes and Infinitely Divisible Distributions. Cambridge University Press, New York (1999)
15. Sethuraman, J.: A constructive definition of Dirichlet priors. Stat. Sinica **4**, 639–650 (1994)
16. Ishwaran, H., James, L.F.: Gibbs sampling methods for stick-breaking priors. J. Am. Stat. Assoc. **96**, 161–173 (2001)
17. Pitman, J.: Combinatorial Stochastic Processes. In: Ecole d'Été de Probabilités de Saint Flour. Lecture Notes in Mathematics, vol. 1875. Springer, Berlin (2006)
18. Blackwell, D., MacQueen, J.B.: Ferguson distributions via Pólya urn schemes. Ann. Stat. **1**, 353–355 (1973)
19. Hoppe, F.: Pólya-like urns and the Ewens' sampling formula. J. Math. Biol. **20**, 91–94 (1984)
20. Antoniak, C.: Mixture of Dirichlet processes with applications to Bayesian nonparametric problems. Ann. Stat. **2**, 1152–1174 (1974)
21. Kingman, J.F.C.: The representation of partition structures. J. Lond. Math. Soc. **18**, 374–380 (1978)
22. Kingman, J.F.C.: Random partitions in population genetics. Proc. R. Soc. Lond. Ser. A. **361**, 1–20 (1978)
23. Lo, A.Y.: Weak convergence for Dirichlet processes. Sankhyā A **45**, 105–111 (1983)

24. Efron, B., Thisted, R.: Estimating the number of unseen species: How many words did Shakespeare Know? Biometrika **63**, 435–447 (1976)
25. Lijoi, A., Mena, R.H., Prünster, I.: Bayesian nonparametric estimation of the probability of discovering new species. Biometrika **94**, 769–786 (2007)
26. Lijoi, A., Prünster, I., Walker, S.G.: Bayesian nonparametric estimators derived from conditional Gibbs structures. Ann. Appl. Probab. **18**, 1519–1547 (2008)
27. Sibuya, M.: Prediction in Ewens-Pitman sampling formula and random samples from number partitions. Ann. Inst. Stat. Math. **66**, 833–864 (2014)
28. Keener, R., Rothman, E., Starr, N.: Distribution of partitions. Ann. Stat. **15**, 1466–1481 (1978)
29. Lijoi, A., Prünster, I.: Models beyond the Dirichlet process. In: Hjort, N.L., et al. (eds.) Bayesian Nonparametrics. Cambridge University Press, Cambridge (2010)
30. Hjort, N.L., et al. (eds.): Bayesian Nonparametrics. Cambridge University Press, Cambridge (2010)
31. Phadia, E.G.: Prior Processes and Their Applications: Nonparametric Bayesian Estimation, 2nd edn. Springer, Switzerland (2013)
32. Escobar, M.D., West, M.: Bayesian density estimation and inference using mixtures. J. Am. Stat. Assoc. **90**, 577–588 (1995)
33. Ishwaran, H., James, F.L.: Generalized weighted Chinese restaurant processes for species sampling mixture models. Stat. Sinica **13**, 1211–1235 (2003)
34. Lee, J., Quintana, F.A., Müller, P., Trippa, L.: Defining predictive probability functions for species sampling models. Stat. Sci. **28**, 209–222 (2013)
35. Pitman, J., Yor, M.: The two-parameter Poisson–Dirichlet distribution derived from a stable subordinator. Ann. Probab. **25**, 855–900 (1997)
36. Carlton, M.A.: Applications of the two-parameter Poisson–Dirichlet distribution. Dissertation, University of California, Los Angeles (1999)
37. James, L.F.: Large sample asymptotics for the two-parameter Poisson–Dirichlet process. Inst. Math. Stat. Collect. **3**, 187–199 (2008)
38. Jang, G.H., Lee, J., Lee, S.: Posterior consistency of species sampling priors. Stat. Sinica **20**, 581–593 (2010)
39. Lo, A.Y.: On a class of Bayesian nonparametric estimates: I. density estimates. Ann. Stat. **12**, 351–357 (1984)
40. Ghosal, S., Ghosh, J.K., Ramamoorthi, R.V.: Posterior consistency of Dirichlet mixtures in density estimation. Ann. Stat. **27**, 143–158 (1999)
41. Lijoi, A., Prüenster, I., Walker, S.G.: On consistency of nonparametric normal mixtures for Bayesian density estimation. J. Am. Stat. Assoc. **100**, 1292–2196 (2005)
42. Gnedin, A., Pitman, J.: Exchangeable Gibbs partitions and Stirling triangles. Zap. Nauchn. Sem. S. POMI **325**, 83–102 (2005). English translation: J. Math. Sci. **138**, 5674–5685 2006)
43. Kerov, S.V: Coherent random allocations, and the Ewens-Pitman formula. Zap. Nauchn. Semi. POMI **325**, 127–145 (1995). English translation: J. Math. Sci. **138**, 5699–5710 (2006)
44. De Iorio, M., Müller, P., Rosner, G.L., MacEachern, S.N.: An ANOVA model for dependent random measures. J. Am. Stat. Assoc. **99**, 205–215 (2004)
45. Gelfand, A., Kottas, A., MacEachern, S.N.: Bayesian nonparametric spatial modeling with Dirichlet process mixing. J. Am. Stat. Assoc. **100**, 1021–1035 (2005)
46. Griffin, J.E., Steel, M.F.J.: Order-based dependent Dirichlet process. J. Am. Stat. Assoc. **101**, 179–194 (2006)
47. Airoldi, E.M., Costa, T., Bassetti, F., Leisen, F., Guindani, M.: Generalized species sampling priors with latent beta reinforcements. J. Am. Stat. Assoc. **109**, 1466–1480 (2014)
48. Jo, S., Lee, J., Müller, P., Quintana, F.A., Trippa, L.: Dependent species sampling models for spatial density estimation. Bayesian Anal. **12**, 379–406 (2017)
49. Griffiths, R.C., Spa?o, D.: Record indices and age-ordered frequencies in exchangeable Gibbs partitions. Electron. J. Probab. **40**, 1101–1130 (2007)
50. Lijoi, A., Mena, R.H., Prünster, I.: Controlling the reinforcement in Bayesian nonparametric mixture models. J. R. Stat. Soc. Ser. B **69**, 715–740 (2007)
51. Feng, S.: The Poisson–Dirichlet Distributions and Related Topics. Springer, Berlin (2010)

52. Perman, M., Pitman, J., Yor, M.: Size-biased sampling of Poisson point processes and excursions. Probab. Theory Relat. Fields **92**, 21–39 (1992)
53. Pitman, J.: Poisson-Kingman partitions. Institute of Mathematical Statistics Lecture Notes-Monograph Series, vol. 40, pp. 245–267 (2003)
54. Pitman, J.: Random discrete distributions invariant under size-biased permutation. Adv. Appl. Probab. **28**, 525–539 (1996)
55. Lijoi, A., Prünster, I., Walker, S.G.: Investigating nonparametric priors with Gibbs structure. Stat. Sinica **18**, 1653–1668 (2008)
56. Lijoi, A., Mena, R.H., Walker, S.G.: Hierarchical mixture modeling with normalized inverse Gaussian priors. J. Am. Stat. Assoc. **100**, 1278–1291 (2005)
57. Ho, M.-W., James, L.F., Lau, J.W. Gibbs partitions (EPPF's) derived from a stable subordinator are Fox H- and Meijer G-transforms. arXiv: 0708.0619

Chapter 5
Methods for Inferences

Abstract This chapter introduces inference methods based on the models presented in previous chapters. We discuss samplers, which are required for hypothesis testing and posterior sampling. After a brief introduction of Markov chain Monte Carlo (MCMC) samplers from A-hypergeometric distributions discussed in Chap. 3, we introduce a direct sampler, which allows us to draw independent samples directly from the target distribution. Gibbs partitions introduced in Chap. 2 and further discussed in Chap. 4 are related to A-hypergeometric distributions of two-rows matrices. We present some interesting topics on samplers from random partitions, including mixing assessment in terms of symmetric functions and construction of direct samplers by simulating stochastic processes on partitions. Finally, aided by information geometry, we discuss maximum likelihood estimation of curved exponential families, which arise in parameterization of the variables of A-hypergeometric distributions.

Keywords A-hypergeometric distribution · Curved exponential family
Direct sampler · Duality of Markov chains · Gibbs partition
Information geometry · Markov chain Monte Carlo · Mixing · Symmetric function

5.1 Sampler

This section introduces samplers from the A-hypergeometric distributions discussed in Sect. 3.2. Suppose that among m categories, $t_i \in [m]$ is the category of the i-th observation of a sample of size n. The count vector $(c_1, \ldots, c_m)$, $c_i := \#\{j; t_j = i\}$ of a sample

$$(t_1, \ldots, t_n) \tag{5.1}$$

should satisfy the homogeneity condition (see Sect. 3.1)

$$c_1 + \cdots + c_m = n. \tag{5.2}$$

If a sequence $(T_1, \ldots, T_n)$ of random variables is n-exchangeable, the count vector follows the A-hypergeometric distribution (Definition 3.4)

© The Author(s) 2018
S. Mano, *Partitions, Hypergeometric Systems, and Dirichlet Processes in Statistics*,
JSS Research Series in Statistics, https://doi.org/10.1007/978-4-431-55888-0_5

$$\mathbb{P}(C_1 = c_1, \ldots, C_m = c_m) = \frac{1}{Z_A(b; x)} \frac{x^c}{c!}. \tag{5.3}$$

Sections 5.1.1 and 5.1.2 introduce MCMC samplers from A-hypergeometric distributions. A Metropolis algorithm is discussed in Sect. 5.1.1, while a direct sampling algorithm is discussed in Sect. 5.1.2.

Section 2.2 introduced the Gibbs partition, which determines the shape of a random Young tableau. The roles in Bayesian nonparametrics were introduced in Sect. 4.5.1. The probability mass function of the size index takes the form

$$\mathbb{P}(C_1 = c_1, \ldots, C_n = c_n) = \frac{v_{n,l(\lambda)} n!}{B_n(v, w)} \frac{x^c}{c!}, \qquad x_i = \frac{w_i}{i!}. \tag{5.4}$$

Here, the categories are the sizes of parts (i.e., n is m in (5.2)). The conditional distribution is an A-hypergeometric distribution of two-row matrix (see Example 3.8). Throughout Sect. 5.1, samplers from random partitions including Gibbs partitions are discussed.

5.1.1 MCMC Samplers

A Metropolis algorithm requires construction of an irreducible and aperiodic Markov chain in the state space (see Sect. 1.2.2). Diaconis and Sturmfels [1] demonstrated the construction using the Gröbner bases. Markov bases are comprehensively introduced in [2].

The state space $\mathscr{F}_b(A) := \{c; Ac = b\}$ is the b-fiber of a matrix A. The set of moves is denoted by $\mathscr{M}(A) := \operatorname{Ker} A \cap \mathbb{Z}^n$, where a move z is written as

$$z = z^+ - z^-, \qquad z_i^+ := \max(z_i, 0), \qquad z_i^- := \max(-z_i, 0).$$

A Markov basis characterizes consecutive moves along the Markov chain.

Definition 5.1 ([1]) If $\mathscr{B} \subset \mathscr{M}(A)$ is a *Markov basis*, for any b and $u, v \in \mathscr{F}_b(A)$, there exist $z_i \in \mathscr{B}$ and $\epsilon_i \in \{-1, 1\}$ such that

$$v = u + \sum_{i=1}^{t} \epsilon_i z_i, \qquad u + \sum_{i=1}^{s} \epsilon_i z_i \in \mathscr{F}_b(A), \qquad \forall s \in [t].$$

The map from a move $z \in \mathscr{M}(A)$ to a binomial $z \mapsto x^{z^+} - x^{z^-}$ is one-to-one. The fundamental theorem of the Markov basis is stated below (a proof is given in Theorem 4.1 of [2]).

Theorem 5.1 ([1]) *A set of moves* $\mathscr{B} = \{z_i; i \in [s]\} \subset \mathscr{M}(A)$ *is a Markov basis if and only if the set of binomials* $\{x^{z_i^+} - x^{z_i^-}; i \in [s]\}$ *generates the toric ideal of the matrix* A.

Example 5.1 (Two-way contingency tables) A two-way $r \times c$ contingency table has $m = r \times c$ categories. The count vector is $(n_{11}, \ldots, n_{1c}, \ldots, n_{r1}, \ldots, n_{rc})$ with sum $n = n_{...}$. Example 3.7 demonstrated that the conditional distribution of counts with fixed marginal sums is the A-hypergeometric distribution. A Markov basis is obvious, consisting of the following moves:

$$z_{ij} = \begin{cases} +1, \ (i, j) = (i_1, j_1), (i_2, j_2), \\ -1, \ (i, j) = (i_1, j_2), (i_2, j_1), \qquad 1 \le i_1 < i_2 \le r, \ 1 \le j_1 < j_2 \le c. \\ 0, \ others, \end{cases}$$

Example 5.2 (*Poisson regression*) In Sect. 1.2.2, we discussed a goodness-of-fit test for the m-level univariate Poisson regression. Example 3.6 showed that the conditional distribution of count vector given the sufficient statistics is the A-hypergeometric distribution. The levels are the categories and the total number of counts in the count vector $(c_1, \ldots, c_m)$ is k (n in (5.2)). Let us consider a Markov basis for the matrix

$$A = \begin{pmatrix} 0 & 1 & \cdots & m-1 \\ 1 & 1 & \cdots & 1 \end{pmatrix}. \tag{5.5}$$

By virtue of Theorem 5.1, a Gröbner basis of the toric ideal is a Markov basis. Therefore, a minimal Gröbner basis given in Proposition 3.2

$$\{e_i + e_j - e_{i+1} - e_{j-1}; 1 \le i < j \le m, i + 2 \le j\} \tag{5.6}$$

is a Markov basis. Simple calculation gives the acceptance ratios in the Metropolis algorithm. The explicit expressions are provided in [3].

Example 5.3 (*Conditional Gibbs partitions*) As shown in Example 3.8, the conditional probability measure of a Gibbs partition with given length is an A-hypergeometric distribution. The sizes of parts are the categories and the count vector $(c_1, \ldots, c_n)$ is the size index (n and k are identical to m and n in (5.2), respectively). The matrix A is given by

$$A = \begin{pmatrix} 0 & 1 & \cdots & n-k \\ 1 & 1 & \cdots & 1 \end{pmatrix}$$

and $b^\top = (n - k, k)$. As the matrix A takes the form of (5.5), its Markov basis is (5.6).

Remark 5.1 Example 5.3 provides clues for construction of an MCMC sampler from (unconditional) Gibbs partitions (5.4) based on a reversible jump MCMC [4]. Green and Richardson [5] discussed such a sampler for posterior sampling from Dirichlet process mixture models. The reversible jump MCMC requires two categories of moves: those that maintain the lengths of partitions, and those that change the length of partitions. The former category is given by the Markov basis (5.6), and the latter category splits one part into two or merges two parts.

5.1.2   Direct Samplers

As discussed in Sect. 1.2.2, a sampler that directly draws independent samples from the target distribution would be beneficial in several respects. Fortunately, such a direct sampler is available for A-hypergeometric distributions, but at the cost of evaluating the normalizing constants (A-hypergeometric polynomials). Methods for evaluating A-hypergeometric polynomials were discussed in Sect. 3.3.

Note that an A-hypergeometric system has the homogeneity constraint (5.2). Equivalently, we have the annihilator

$$\sum_{i=1}^{m} \theta_i - n.$$

Applying this annihilator to the A-hypergeometric polynomial provides a relation among the A-hypergeometric polynomials

$$\sum_{i=1}^{m} x_i Z_A(b - a_i; x) = n Z_A(b; x).$$

Note that the degree of the polynomial $Z_A(b - a_i; x)$ is $n - 1$. This expression can be probabilistically interpreted as follows. Consider a Markov chain with a state space consisting of A-hypergeometric polynomials. Along each step of the chain, the degree of the polynomial decreases by one. The transition probability from the polynomial $Z_A(b; x)$ to the polynomial $Z_A(b - a_i; x)$ is given by

$$\frac{Z_A(b - a_i; x)}{Z_A(b; x)} \frac{x_i}{n} = \frac{\mathbb{E}(C_i)}{n} =: p(i), \tag{5.7}$$

which coincides with the expected proportion of the i-th category. Likewise, define

$$p(i|j_1, \ldots, j_{l-1}) := \frac{Z_A(b - a_{j_1} - \cdots - a_{j_{l-1}} - a_i; x)}{Z_A(b - a_{j_1} - \cdots - a_{j_{l-1}}; x)} \frac{x_i}{n - l + 1}, \tag{5.8}$$

if $b \geq a_{j_1} + \cdots + a_{j_{l-1}} + a_i$, and set $p(i|j_1, \ldots, j_{l-1}) = 0$ otherwise. This is the expected proportion of the i-th category after removing $j_1, \ldots, j_{l-1}$ categories from the sufficient statistics b (see Sect. 3.2). A sample path of this Markov chain is a draw of categories (5.1) from the A-hypergeometric distribution (5.3). The procedure is implemented by the following sequential sampling algorithm.

Algorithm 5.1 ([3]) *Sequential sampling of a vector of categories* $(t_1, \ldots, t_n)$, *when the count vector* $(c_1, \ldots, c_m)$, *follows the A-hypergeometric distribution* (5.3).

1. *Pick $t_1 = i$ with probability $p(i)$ defined by* (5.7).
2. *For $l = 2, \ldots, n$, pick $t_l = i$ with probability $p(i|t_1, \ldots, t_{l-1})$ defined by* (5.8).

This algorithm is now demonstrated with examples.

Example 5.4 (*Two-way contingency tables*) Algorithm 5.1 provides a direct sampler from a two-way contingency table. The MCMC sampler was discussed in Example 5.1. By sequentially picking a count of one cell in the contingency table, we can sample a two-way contingency table with fixed marginal sums. The step-by-step implementation of this algorithm is best shown by a toy example. Let $b = (2, 1, 2, 1)$. The sample path $t = (2, 3, 1)$ corresponds to a sequence of contingency tables

$$
\begin{array}{cc|c}
1 & 1 & 2 \\
1 & 0 & 1 \\
\hline
2 & 1 & 3
\end{array}
\rightarrow
\begin{array}{cc|c}
1 & 0 & 1 \\
1 & 0 & 1 \\
\hline
2 & 0 & 2
\end{array}
\rightarrow
\begin{array}{cc|c}
1 & 0 & 1 \\
0 & 0 & 0 \\
\hline
1 & 0 & 1
\end{array}
\rightarrow 0,
$$

or a sequence of monomials $x_1 x_2 x_3 \rightarrow x_1 x_3 \rightarrow x_1$. The probability of generating this path is

$$
\frac{x_1 x_3}{x_1 x_2 x_3 + x_1^2 x_4/2!} \frac{x_2}{3} \times \frac{x_1}{x_1 x_3} \frac{x_3}{2} = \frac{x_1 x_2 x_3}{x_1 x_2 x_3 + x_1^2 x_4/2!} \frac{1}{3!} = \frac{1}{3!} \frac{x_1 x_2 x_3}{Z_A(b; x)}.
$$

Here, the factor $1/3!$ appears because there are $3!$ equiprobable paths for generating the count vector $c = (1, 1, 1, 0)$. As another table is possible, the denominator of the first fraction is binomial. As an example of real-life data, Table 5.1 shows a partial result of the 13th study of the Japanese national character in 2013, a regular survey carried out by the Institute of Statistical Mathematics since 1953 [6]. The table tabulates the answers to Question 2.11b by respondents in the 20–29 years age bracket. The question consists of two items. The "experience" item represents "I want to experience as much as possible in order to develop my capacities in work and play", and the "avoid trouble" item represents "I want to avoid trouble as much as possible and live in peace and quiet". The chi-squared value is 1.561 and the p-value of the Fisher's exact test of rows-and-columns independence (computed by directly enumerating all possible tables) is 0.233. The estimated p-value based on 1,000 samples from the direct sampler was 0.236. The direct sampler performed well, despite the small number of samples, as confirmed by the similarity between the true and estimated probabilities shown in Fig. 5.1. The computation was implemented by Risa/Asir version 20160405 [7] with the gtt_ekn.rr package [8, 9].

Table 5.1 A partial result of the 13th study of the Japanese national character carried out in 2013

	Experience	Avoid trouble	
Males	53	19	72
Females	56	31	87
	109	50	159

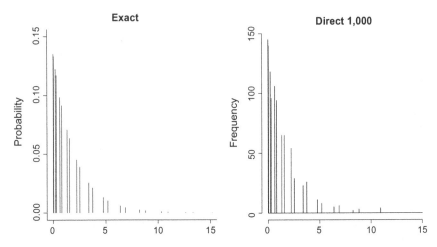

Fig. 5.1 Probabilities of tables those chi-squared values are in the horizontal axis. Left is the true probabilities. Right is the number of tables among 1,000 tables generated by the direct sampler

Example 5.5 (*Poisson regression* [10]) In Example 1.2, we introduced a goodness-of-fit test for the univariate Poisson regression. The MCMC sampler was discussed in Example 5.2. With the aid of Algorithm 5.1, we can sample a vector of levels by sequentially picking a level among [m]. For the data in Example 1.2, the p-value of the goodness-of-fit of the Poisson regression was estimated. First, following Diaconis et al. [10], we estimated the p-value by the MCMC sampler. Based on the 90, 000 steps with the initial 10, 000 steps having been discarded as the burn-in, the estimated p-value was 0.023. The issue here is whether 10, 000 steps is sufficient to be free from possible bias comes from departure from the stationarity. The estimate with direct sampler can give an answer to this problem, since the estimate should have no bias and the Monte Carlo error can be diminished simply by running a long chain. The estimate based on 900, 000 samples from the direct sampler was 0.026. Since this estimate is close to the estimate by the MCMC sampler, we can conclude that the MCMC sampler gave a reasonable estimate, as confirmed by the similarity between the two histograms in Fig. 5.2.

Example 5.6 (*Conditional Gibbs partitions*) In Example 5.3, we discussed the Metropolis algorithm from the conditional probability measure of a Gibbs partition given length. Similarly, to the direct sampler for the Poisson regression in Example 5.5, we can sample a partition from the conditional probability measure by sequentially picking single row of the Young tableau. For a given length, Stewart [11] proposed an algorithm that samples from the conditional probability measure of the Ewens sampling formula (2.13). The MCMC sampler from the (unconditional) Gibbs partition was discussed in Remark 5.1. A Gibbs partition (5.4) can be directly sampled by selecting the number of rows of the Young tableaux (length of partition)

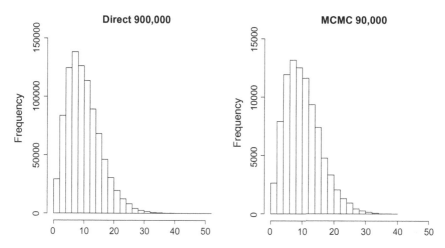

Fig. 5.2 Histograms of the chi-squared values. Left is obtained by the direct sampler with 900,000 samples. Right is obtained by the MCMC sampler based on 90,000 steps with the initial 10,000 steps been discarded

k with probability in (2.22), then sampling the lengths of the rows $(n_1, \ldots, n_k)$ by using Algorithm 5.1.

Remark 5.2 In Sect. 4.5.1, we introduced an analog of Blackwell–MacQueen's urn scheme (Algorithm 4.2) as a sampler from Gibbs partitions. Although being a direct sampler, the urn scheme is available if and only if the Gibbs partition is infinitely exchangeable (see Sect. 1.1 for the definition). Some Gibbs partitions discussed so far are not infinitely exchangeable (see Remark 4.12). The direct sampler introduced in Example 5.6 works without infinite exchangeability.

5.1.3 Mixing and Symmetric Functions

Given a target distribution, one can construct a Metropolis sampler by a Markov basis. The mixing depends on the chosen Markov basis. A sampler from a Gibbs partition admits any Markov basis of the fiber

$$(c_1, \ldots, c_n), \qquad 1 \cdot c_1 + 2 \cdot c_2 + \cdots + n \cdot c_n = n. \tag{5.9}$$

The Metropolis sampler mentioned in Remark 5.1 utilizes the Markov basis (5.6) of the matrix (5.5) and the splitting and merging moves

$$\{e_i + e_j - e_{i+j}; 1 \le i, 1 \le j, i + j \le n\}. \tag{5.10}$$

It can be shown that the moves (5.10) form a Markov basis of (5.9). In fact, Hanlon [12] considered a Metropolis sampler from the Ewens sampling formula (2.12) with moves (5.10) only. He discussed the mixing using properties of Jack symmetric functions (the symmetric functions appear here are briefly introduced in Sect. A.1). Diaconis and Lam [13] discussed a MCMC sampler from the Gibbs partition associated with Macdonald symmetric functions (see Example 2.15). The employed moves are different from (5.10). The following algorithm utilizes auxiliary variables, and is known as the block spin algorithm in statistical mechanics [14].

Algorithm 5.2 ([13]) *Sampling of a Gibbs partitions associated with Macdonald symmetric functions (Examples 2.15) of weight n.*

1. *Set $t = 0$ and pick an initial sample $\lambda^{(0)}$.*
2. *Pick parts σ with probability*

$$\mathbb{P}(\Sigma = \sigma) = \frac{1}{q^n - 1} \prod_{i=1}^{n} \left(\frac{c_i(\lambda^{(t)})}{c_i(\lambda^{(t)} \backslash \sigma)} \right) (q^i - 1)^{c_i(\sigma)}.$$

3. *Pick parts $\sigma' \vdash |\sigma|$ with probability*

$$\mathbb{P}(\Sigma' = \sigma') = \frac{t}{t-1} \prod_{i=1}^{n} \left\{ \frac{1}{i} \left(1 - \frac{1}{t^i} \right) \right\}^{c_i(\sigma')} \frac{1}{c_i(\sigma')!}.$$

4. *Set $\lambda^{(t+1)} = (\lambda^{(t)} \backslash \sigma) \cup \sigma'$, increment t to $t+1$, and return to Step 2.*

Remark 5.3 In each iteration of this algorithm, parts σ are replaced with parts σ', while retaining $\lambda \backslash \sigma$. In this step, the parts $\lambda \backslash \sigma$ are the auxiliary variables.

Consideration of the transition probability matrix involves representations of Macdonald symmetric functions. As shown in Theorem 3.1 of [13], the eigenvalues are given by

$$\beta_\lambda = \frac{t}{q^n - 1} \sum_{i=1}^{l(\lambda)} \frac{q^{\lambda_i} - 1}{t^i}$$

and the eigenvectors are

$$f_\lambda(\rho) = X_\rho^\lambda(q, t) \prod_{i=1}^{l(\rho)} (1 - q^{\rho_i}), \qquad X_\rho^\lambda(q, t) := \sum_\mu \chi_\rho^\mu K_{\mu\lambda}(q, t).$$

Here, $X_\rho^\lambda(q, t)$ are the coefficients occurring in the expansion of Macdonald symmetric functions in terms of power sums, where $K_{\lambda,\rho}(q, t)$ are the two-parameter Kostka numbers and $\chi_\rho^\mu = \langle s_\mu, p_\rho \rangle$ (see Sect. VI.8 of [15]). For a chain beginning at the partition (n), the chi-squared distance after t steps is

$$\chi^2_{(n)}(t) = \sum_{\lambda} \frac{\{(P^t)_{(n)\lambda} - \mu_n(\lambda)\}^2}{\mu_n(\lambda)} = \sum_{\lambda \neq (n)} \beta_\lambda^{2t} \bar{f}_\lambda^2(n),$$

where the eigenvectors $\bar{f}_\lambda(n)$ are normalized to have unity norm. Theorem 5.1 of [13] provides the explicit form of the upper bound.

5.1.4 Samplers with Processes on Partitions

In Sects. 5.1.1 and 5.1.3, MCMC samplers for random partitions were discussed. In Sect. 5.1.2, direct sampling from random partitions using the direct sampler for the A-hypergeometric distribution was discussed. In this section, we introduce a method which enables direct sampling of random partitions by simulating random graphs generated by processes on partitions. The notion of *dualities* between processes on partitions and measure-valued diffusions are the key. The background is briefly explained in Sect. A.2. Such inference methods have been developed in probabilistic studies on genetic diversity in the field of population genetics. An extensive survey of inferences in population genetics is [16].

The Ewens sampling formula satisfies the recurrence relation (A.12), which can be understood with *Kingman's coalescent*. Although the recurrence relation is a linear system, it is extremely high-dimensional. It is unrealistic to solve it exactly. Therefore, we consider a stochastic algorithm to sample a partition. Such an algorithm was proposed by Griffiths and Tavaré [17]. Solving high-dimensional system of linear equations via stochastic simulations was originated by von Neumann and Ulam, and extended by Forsythe and Leibler [18].

Kingman's coalescent is a process to generate random trees. Set the time direction from the root to the leaves with the origin at the leaves. Let x_t denote the state of partition at time t. The waiting time $(-T) > -\infty$ exists such that $x_{-T} = e_1$. The recurrence relation has the form of a forward equation

$$p(x_t) = \sum_{\{x_{t-1}\}} p(x_t|x_{t-1})p(x_{t-1}).$$

The problem is to compute the probability $p(x_0)$ of a sample x_0. Since we know the forward transition probabilities $p(x_t|x_{t-1})$, it is straightforward to generate a tree from the root e_1 to the leaves. However, the probability that a sample path hits the state x_0 is extremely low. A possible alternative is generating a tree from x_0 backward in time and stopping at time $(-T)$. A difficulty associated with this method is that we do not know the backward transition probabilities $p(x_{t-1}|x_t)$. From Bayes' rule the backward transition probability can be expressed as $p(x_{t-1}|x_t) = p(x_t|x_{t-1})p(x_{t-1})/p(x_t)$; however, it is unknown as we do not know $p(x_{t-1})/p(x_t)$. The following importance sampling representation is based on an approximation of the backward transition probabilities $\hat{p}(x_{t-1}|x_t)$:

$$p(x_0) = \mathbb{E}_{\hat{p}}\left[\frac{p(x_0|x_{-1})}{\hat{p}(x_{-1}|x_0)} \cdots \frac{p(x_{-T+1}|x_{-T})}{\hat{p}(x_{-T}|x_{-T+1})} p(x_{-T})\right]$$

$$\approx \frac{1}{N} \sum_{i=1}^{N} \left\{ \prod_{j=0}^{-T+1} \frac{p(x_j^{(i)}|x_{j-1}^{(i)})}{\hat{p}(x_{j-1}^{(i)}|x_j^{(i)})} p(x_{-T}) \right\},$$

where $\mathbb{E}_{\hat{p}}$ denotes expectation taken over backward paths $\{x_{-1}, \ldots, x_{-T}\}$ with the approximated backward transition probabilities, and $x^{(1)}, \ldots, x^{(N)}$ are the N independent sample paths. If $\hat{p}(x_{t-1}|x_t) = p(x_{t-1}|x_t)$, the importance weight becomes $p(x_0)$. Stephens and Donnelly [19] proposed following approximation procedure, which was formalized by de Iorio and Griffiths [20].

Let us consider a sampler from the symmetric m-variate Dirichlet-multinomial distribution as an example. Let the sample be $n = (n_1, \ldots, n_m)$. In Sect. 4.3, we observed that $p(n) = \langle q_n, \pi_\alpha \rangle$, where q_n is defined in (A.10) and π_α is the density of the Dirichlet distribution. However, suppose that we do not know these expressions and want to approximate $p(n)$. The exchangeability assumption in sequential sampling demands that the approximated probability $\hat{p}(n)$ satisfies

$$\pi(i|n - e_i)\hat{p}(n - e_i) = \frac{n_i}{n}\hat{p}(n), \tag{5.11}$$

where $\pi(i|n - e_i)$ is the probability that an additional type chosen is of type i, given a sample of $n - e_i$. To determine π and $\hat{p}$, de Iorio and Griffiths [20] proposed using the condition for the generator of a diffusion (A.7):

$$\left\langle L_j \frac{\partial q_n}{\partial x_j}, \pi_\alpha \right\rangle = 0, \tag{5.12}$$

where L_j is

$$L_j := \sum_i \frac{1}{2}a_{ij}(x)\frac{\partial}{\partial x_i} + b_j(x).$$

The condition (5.12) yields

$$n_j(n - 1 + m\alpha)\hat{p}(n) = n(n_j - 1)\hat{p}(n - e_j) + \alpha \sum_{i=1}^{m}(n_i + 1 - \delta_{ij})\hat{p}(n + e_i - e_j).$$

Using (5.11), we have

$$\pi(i|n) = \frac{\alpha + n_i}{m\alpha + n}.$$

Note that this expression coincides with the prediction rule of the Dirichlet-multinomial distribution (4.11). Therefore, $\hat{p}(n) = p(n)$. This coincidence occurs because the Wright–Fisher diffusion (A.9) is reversible (see Sect. A.2). This approx-

Table 5.2 The number of G/C copies among 20 gene copies at the 131 variable sites

G/C copies	20	19	18	17	16	15	14	13	11	10	8	7	4	3	1
Sites	39	48	17	10	5	2	2	1	1	1	1	1	1	1	1

imation procedure is useful for models that do not have exchangeability and/or reversibility.

In the importance sampler introduced here, simulation of random trees stops at the root at time $(-T) < -\infty$, and the boundary condition $p(x_{-T})$ is applied. However, in simulating general random graphs including those in Example 5.7, we encounter a problem when we can stop simulation of random graphs. This is a problem called *coupling from the past*. Propp and Willson [21] discussed the problem in general MCMC setup. A path from the infinite past can be considered to be a path from the stationary distribution of the sampler. Therefore, if a path from a finite past couples with a path from the infinite past, the path can be considered to be a path from the stationary distribution. The simulation can be stopped at the time of coupling, because events before the coupling have no influence upon the sample. Following this principle, we can stop generation of random graphs when coupling occurs. In a tree structure, coupling must occur by $(-T)$. Refer to [22] for an application to a branching-coalescent.

Example 5.7 (*Voter model*) The following voter model was considered in [23]. Refer to Part II of [24] for the background of voter models. Consider n groups of sites and each group comprises N sites. All sites are connected to each other. Each site has one of the two types of opinions. With fixed rates, a randomly chosen neighbor of a site x follows the opinion of x. If the neighbor is in the same group, the rate does not depend on the opinion of x. If the neighbor is in a different group, the rate depends on the type of the opinion of x. Each site flips its opinion randomly with a fixed rate. In the diffusion limit $N \to \infty$, a Wright–Fisher diffusion on $[0, 1]^n$ appears. In [23], an importance sampler was constructed for the maximum likelihood estimation of the bias by simulating random graphs following a branching-coalescent. Although the bias breaks reversibility, the approximation procedure discussed above was useful to construct an efficient sampler. In genetics, it is known that there are family of genes among which a part of gene is 'copied and pasted' to a part of another gene. The copy and past mechanism is called gene conversion. It is also known that G/C nucleotides are more frequently copied and pasted than A/T nucleotides. The biased voter model introduced here was applied to a mouse gene family data set which consists of $n = 20$ genes with 131 variable nucleotide sites with assuming that each site follows the biased voter model independently. The data is tabulated in Table 5.2. It was estimated that G/C nucleotides are copied and pasted 1.2 times more frequently than A/T nucleotides. The measure-valued diffusion with infinitely-many type of opinions without bias was studied by Shimizu [25]. He found that the Ewens sampling formula appears. See [25] for the details.

5.2   Estimation with Information Geometry

In statistical point of view, assuming variables of an A-hypergeometric distribution
to be parameterized by small number of parameters is reasonable to avoid overparameterization. Moreover, for two-row matrices, the maximum likelihood estimator (MLE) does not exist for a count vector (Theorem 3.7). This section discusses
maximum likelihood estimation of curved exponential families, which appear by
parameterizing variables of A-hypergeometric distributions. Information geometry
[26] is a powerful tool to discuss the issue. Assume that the sufficient statistics c
of an A-hypergeometric distribution is average of N count vectors (see Sect. 3.2.1).
Then, c is a point in the Newton polytope $\mathrm{New}(Z_A(b; x))$. The log likelihood is

$$N\{\xi^i c_i - \psi(\xi)\}, \qquad i \in [m], \tag{5.13}$$

where $\xi^i = \log x_i \in \mathbb{R}$ and $\psi(\xi) = \log Z_A(b; x)$. Einstein's summation convention
will be used hereafter; indices denoted by a repeated letter, where the one appears
as a superscript while the other appears as a subscript, are summed up. The moment
map $\mathbb{E}(C) : \mathbb{R}^m / \mathrm{Im} A^\top \ni \log y \mapsto \eta$ in (3.2.1) provides the dual coordinate system
in the sense of information geometry. The dual coordinate is in (3.21) and the Fisher
metric is

$$g_{ij} := \partial_i \partial_j \psi(\xi) = \frac{Z_A(b - a_i - a_j; x)}{Z_A(b; x)} x_i x_j I_{\{b-a_i-a_j \geq 0\}} - \eta_i \eta_j + \eta_i \delta_{i,j},$$

respectively. Because of the dually flatness of the exponential family, the full exponential family is e-flat and also m-flat [26].

Let M be a submanifold of $\mathrm{New}(Z_A(b; x))$ on which the curved exponential
family is defined. Let the coordinate system of M be u^a, $a \in [l]$. We will use the
dual coordinate system $\eta(u)$ to represent a point in $\mathrm{New}(Z_A(b; x))$. An estimator is
a mapping from $\mathrm{New}(Z_A(b; x))$ M:

$$f : \mathrm{New}(Z_A(b; x)) \to M, \qquad c \mapsto \hat{u} = f(c).$$

Let us call the inverse of the estimator $A(u) = f^{-1}(u)$ the estimating manifold
corresponding to the point $u \in M$. Let us prepare a new coordinate system of
$\mathrm{New}(Z_A(b; x))$ around $\eta(u)$: A point η is indexed by (u, v), where v is the index
of η in $A(u)$, where $\eta(u) = \eta(u, 0)$. The tangent space of M at $\eta(u)$ is spanned by
∂_a, while the tangent space of $A(u)$ is spanned by $\partial_\kappa, l + 1 \leq \kappa \leq m$. The following
theorem is fundamental.

Theorem 5.2 ([26]) *For a curved exponential family with submanifold M, an
estimator $\hat{u}$ is consistent if and only if the estimating submanifold A contains
point $\eta(u)$ as $N \to \infty$. The asymptotic covariance matrix of the estimator sat-
isfies* $\lim_{N \to \infty} N\mathbb{E}[(\hat{u}^a - u^a)(\hat{u}^b - u^b)] = \bar{g}^{ab}$, *where* $\bar{g}^{ab} := (g_{ab} - g_{a\kappa} g^{\kappa\lambda} g_{b\lambda})^{-1}$.

The estimator is first-order asymptotically efficient if and only if $A(u)$ and M are orthogonal.

Let us concentrate on a specific example of curved exponential family, which appears by parameterizing variables of the A-hypergeometric distribution in Example 3.8 with $x_i = (1 - \alpha)_{i-1}/i!$, $\alpha < 1$. The likelihood is given in (2.29). This curved exponential family appears as the conditional distribution of an infinite exchangeable Gibbs partition given length, which was discussed in Sect. 4.5.1. The submanifold M is now a curve parameterized by the parameter α. The generalized odds ratios (3.22) become

$$y_i = \frac{2^{i+1}}{(i+2)!} \frac{(1-\alpha)_{i+1}}{(1-\alpha)^{i+1}}, \qquad i \in [n-k-1]. \tag{5.14}$$

The image of the moment map M is now a smooth open curve in the relative interior of $\mathrm{New}(Z_A(b; x))$. One of the limit points is $\eta = (k-1)e_1 + e_{n-k+1}$, which appears as $\alpha \to 1$. This is a vertex of $\mathrm{New}(Z_A(b; x))$ and the Fisher metric is zero. Another limit point appears as $\alpha \to -\infty$, where

$$\eta_i = \binom{n}{i} \frac{S(n-i, k-1)}{S(n, k)}$$

and the Fisher metric is

$$g_{ij} = \binom{n}{i, j} \frac{S(n-i-j, k-2)}{S(n, k)} I_{\{n-k+2 \geq i+j\}} - \eta_i \eta_j + \eta_i \delta_{i,j}.$$

No MLE exists if c is in the normal fan of M at $\alpha = -\infty$. The inverse of the N times the asymptotic variance is $g_{\alpha\alpha} = \|\partial_\alpha^2\| = g_{ij}\partial_\alpha\xi^i\partial_\alpha\xi^j$, where $\partial_\alpha\xi^i = \sum_{j=1}^{i-1}(\alpha - j)^{-1}, i \geq 2$, and $\partial_\alpha\xi^1 = 0$, which is the squared norm of the tangent vector along with the curve M. The squared norm vanishes as $\alpha \to 1$ and diverges as $\alpha \to -\infty$, which implies that the model is singular at these limit points.

Example 5.8 When $n = k + 3 \geq 6$, the Newton polytope $\mathrm{New}(Z_A(b; x))$ is the convex hull of the three vertices $(n-6, 3, 0, 0)^\top$, $(n-5, 1, 1, 0)^\top$, and $(n-4, 0, 0, 1)^\top$. The image of the moment map is

$$\begin{pmatrix} \eta_1 \\ \eta_2 \\ \eta_3 \\ \eta_4 \end{pmatrix} = \begin{pmatrix} n-6 \\ 3 \\ 0 \\ 0 \end{pmatrix} + \frac{\frac{3!}{n-5}y_1}{1 + \frac{3!}{n-5}y_1 + \frac{3!}{(n-4)(n-5)}y_2} \begin{pmatrix} 1 \\ -2 \\ 1 \\ 0 \end{pmatrix}$$

$$+ \frac{\frac{3!}{(n-4)(n-5)}y_2}{1 + \frac{3!}{n-5}y_1 + \frac{3!}{(n-4)(n-5)}y_2} \begin{pmatrix} 2 \\ -3 \\ 0 \\ 1 \end{pmatrix}.$$

One of the limit points of the curve M with $\alpha \to 1$ is $(n-4, 0, 0, 1)^\top$, while the other limit point with $\alpha \to -\infty$ is

$$\left(\frac{(n-4)^2}{n-2}, \frac{(3n-11)(n-4)}{(n-2)(n-3)}, \frac{4(n-4)}{(n-2)(n-3)}, \frac{2}{(n-2)(n-3)} \right)^\top.$$

The latter point is in the relative interior of New$(Z_A(b; x))$, but in the limit $n \to \infty$ it tends to $(n-6, 3, 0, 0)$, which is a vertex of New$(Z_A(b; x))$. An analysis of the estimating equation tells us that MLE does not exist for small n. It can be shown [3] that the MLE exists uniquely if and only if

$$c_3 + 3c_4 > \frac{2(2n-5)}{(n-2)(n-3)}. \tag{5.15}$$

Let us see (5.15) is certainly the condition of the orthogonal projection around $\alpha \to -\infty$. Let $B_{\alpha i} := \partial_\alpha \eta_i(-\infty, 0) = g_{ij}\partial_\alpha \xi^j$ and $B_{\kappa i} := \partial_\kappa \eta_i(-\infty, 0)$, where $\partial_\alpha = B_{\alpha i}\partial^i$ and $\partial_\kappa = B_{\kappa i}\partial^i$ are the tangent vectors of M and $A(-\infty)$ expressed in terms of basis $\{\partial^i\}$, respectively. Taking $\partial_\kappa = \delta_{\kappa 2}(c_i - \eta_i(u))\partial^i$, the condition of possibility for the orthogonal projection is

$$g_{\alpha 2} = \langle \partial_\alpha, \partial_2 \rangle = B_{\alpha i}B_{2j}g^{ij} = \partial_\alpha \xi^j (c_j - \eta_j(-\alpha)) > 0,$$

which is equivalent to (5.15). If the MLE exists, the asymptotic variance with $N \to \infty$ is $g^{\alpha\alpha}/N \sim n(\alpha-1)^3(\alpha-2)/(4N)$ for large n. The asymptotic variance increases linearly with sample size n. Figure 5.3 depicts the projection of the Newton polytope for $n = 10$ and $k = 7$ onto the η_3-η_4 plane, which is the lower triangle of the diagonal, and the submanifold M is the curve. The estimating manifold for the case of $c =$

Fig. 5.3 The Newton polytope for $n = 10$ and $k = 7$ projected onto the η_3-η_4 plane is the lower triangle. See text for the MLE on it

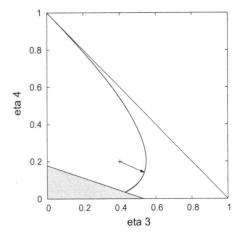

(4.8, 1.6, 0.4, 0.2) is shown by the arrow, and the MLE is $\hat{\alpha} = 0.073$. If c is in the shaded region, which is the normal fan of M at $\alpha = -\infty$, no MLE exists.

Essentially the same argument provides a classical result on the existence of the MLE for a sample taken from the Dirichlet distribution. The log-likelihood of the m-variate symmetric Dirichlet-multinomial distribution of parameter $(-\alpha) > 0$ is given by (5.13) with $\psi(\xi) = \log(-m\alpha)_n$, $x_i = (-\alpha)_i / i!$. This is a curved exponential family. Using the variation-diminishing property of the Laplace transform, Levin et al. [27] (Theorem 1) proved that the MLE exists uniquely if and only if

$$\sum_{i=1}^{n} i^2 c_i > n + \frac{n(n-1)}{m} \tag{5.16}$$

is satisfied. In our context, the assertion is as follows. The image of the moment map is now the partition polytope P_n instead of the Newton polytope. The submanifold M is parameterized by $\alpha < 0$ and the two limit points are $\eta = e_n$ and

$$\eta_i = \frac{(m-1)^{n-i}}{m^{n-1}} \binom{n}{i},$$

which correspond to limits of $\alpha \to 0$ and $\alpha \to -\infty$, respectively. The MLE does not exist if c is in the normal fan at $\alpha = -\infty$, which is equivalent to (5.16).

Remark 5.4 Keener et al. [28] discussed several issues around parameter estimation of the Dirichlet-multinomial distribution. The Dirichlet-multinomial distribution is a Pitman partition of parameter $\alpha < 0$ and $\theta = -m\alpha$, $m \in \mathbb{Z}_{\geq 1}$ (see Example 4.9). Hoshino [29] discussed the maximum likelihood estimation of the Pitman partition.

Before closing this section, let us briefly illustrate numerical evaluation of the MLE. We concentrate here on the A-hypergeometric distribution of two-row matrix A. Methods for general A-hypergeometric distributions are given in [30]. For the full exponential family (5.13), the MLE is

$$\hat{y} := \mathrm{argmax}_y f(y), \qquad f(y) = \sum_{i=1}^{n-k-1} c_{i+2} \log y_i - \psi(y).$$

The derivative is

$$\frac{\partial f}{\partial y_i} = y_i^{-1}(c_{i+2} - \eta_{i+2}(y)), \qquad i \in [n-k-1].$$

Evaluate $\hat{y}$ is equivalent to finding the inverse image of the map $c = \eta(\hat{y})$. A simple gradient descent algorithm is as follows.

Table 5.3 Word frequency spectra of *Alice in Wonderland* (left) and *Through the looking-glass* (right). Entries for $i > 30$ are omitted

i	s_i	η_i	i	s_i	η_i	i	s_i	η_i
1	1491	1579.94	11	26	26.71	21	7	9.73
2	460	410.76	12	30	23.33	22	9	9.04
3	259	207.59	13	22	20.60	23	2	8.42
4	148	130.38	14	19	18.35	24	3	7.87
5	113	91.48	15	12	16.48	25	1	7.38
6	78	68.68	16	21	14.90	26	5	6.93
7	61	53.97	17	12	13.55	27	3	6.53
8	47	43.83	18	11	12.39	28	7	6.16
9	28	36.49	19	16	11.38	29	5	5.82
10	26	30.98	20	9	10.50	30	2	5.52

Algorithm 5.3 *Finding inverse image of the map $c = \eta(\hat{y})$.*

1. *Set $i = 0$ and take small $\epsilon > 0$. Provide $y^{(0)}$ and set $\eta^{(0)} = \eta(y^{(0)})$.*
2. *End if*

$$\frac{\partial f^{(i)}}{\partial y_j} = (y_j^{(i)})^{-1}(c_{j+2} - \eta_{j+2}^{(i)}) \approx 0, \qquad j \in [n - k - 1].$$

3. *Else set*

$$y_j^{(i+1)} = y_j^{(i)} + \epsilon \frac{\partial f^{(i)}}{\partial y_j}, \qquad \eta^{(i+1)} = \eta(y^{(i+1)})$$

increment i to $i + 1$, and return to Step 2.

If we use Newton's method, which is called the natural gradient method in information geometry, $\partial f/\partial y_i$ in Step 2 of Algorithm 5.3 is replaced with

$$\sum_{j=1}^{n-k-1} (H^{-1})_{ij}(c_{j+2} - \eta_{j+2}(y)), \qquad (H)_{ij} := \frac{\partial \eta_{i+2}}{\partial y_j} = y_j^{-1} g_{i+2,j+2}.$$

Compared with the simple gradient descent algorithm, Newton's method demands cost of the matrix inversion. For the curved exponential family, slight modification to the algorithm is needed. The algorithm for the parameterization (5.14) is given in [3].

Example 5.9 (Conditional Gibbs partition) The data sets considered are from [31] and concern word usage of Lewis Carroll in two works, namely, *Alice's Adventure in Wonderland (Alice in Wonderland)* and *Through the looking-glass and what Alice found there (Through the looking-glass)*. An empirical Bayes approach is as follows. In these data, the size index c_i is the number of word types that occur exactly i

times. *Alice in Wonderland* consists of $n = 26, 505$ word tokens, and the number of different word types in the full text of $26, 505$ word tokens is $k = 2, 651$. For example, a word type "Alice" occurs exactly 386 times and other word types do not occur exactly 386 times, so $c_{386} = 1$. Consider application of a Gibbs partition given the length $k = 2, 651$. The conditional MLE of α was carried out with the A-hypergeometric distribution. To evaluate the A-hypergeometric polynomials, the asymptotic approximation (Lemma 2.1) was employed. After 56 iterations of the gradient descent, $\hat{\alpha}$ almost converged to 0.441. For *Through the looking-glass*, $n = 28, 767, k = 3, 085$, and $\hat{\alpha} = 0.478$. The latter is Carroll's second story about Alice. We might hypothesize that Carroll benefited from his experience in writing *Alice in Wonderland*, and that *Through the looking-glass* might be characterized by the greater vocabulary richness [31]. This hypothesis is concordant with the result here, because larger α implies stronger tendency to use word type that have never occurred. Table 5.3 displays word frequency spectra of *Alice in Wonderland* and *Through the looking-glass*.

References

1. Diaconis, P., Sturmfels, B.: Algebraic algorithms for sampling from conditional distributions. Ann. Statist. **26**, 363–397 (1998)
2. Aoki, S., Hara, H., Takemura, A.: Markov Bases in Algebraic Statistics. Springer, New York (2012)
3. Mano, S.: Partition structure and the A-hypergeometric distribution associated with the rational normal curve. Electron. J. Stat. **11**, 4452–4487 (2017)
4. Green, P.J.: Reversible jump Markov chain Monte Carlo computation and Bayesian model determination. Biometrika **82**, 711–732 (1995)
5. Green, P.J., Richardson, S.: Modelling heterogeneity with and without the Dirichlet process. Scand. J. Statist. **28**, 355–375 (2001)
6. Study on the Japanese National Character. http://www.ism.ac.jp/kokuminsei/en/table/index.htm
7. Risa/Asir (Kobe distribution) Download Page. http://www.math.kobe-u.ac.jp/Asir/asir.html
8. Goto, Y., Tachibana, Y., Takayama, N.: Implementation of difference holonomic gradient methods for two-way contingency table. Comput. Algebra Relat. Top. RIMS Kôkyûroku **2054**, 11 (2016)
9. Tachibana, Y., Goto, Y., Koyama, T., Takayama, N.: Holonomic gradient method for two way contingency tables. arXiv: 1803.04170
10. Diaconis, P., Eisenbud, B., Sturmfels, B.: Lattice walks and primary decomposition. In: Sagan, B.E., Stanley, R.P. (eds.) Mathematical Essays in Honor of Gian-Carlo Rota (Cambridge, MA, 1996). Progress in Mathematics, vol. 161, pp. 173–193. Birkhäuser, Boston (1998)
11. Stewart, F.M.: Computer algorithm for obtaining a random set of allele frequencies for a locus in an equilibrium population. Genetics **86**, 482–483 (1997)
12. Hanlon, P.: A Markov chain on the symmetric group and Jack symmetric functions. Discret. Math. **90**, 123–140 (1992)
13. Diaconis, P., Lam, A.: A probabilistic interpretation of the Macdonald polynomials. Ann. Probab. **40**, 1861–1896 (2012)
14. Newman, M.E.J., Barkema, G.T.: Monte Carlo Methods in Statistical Physics. Clarendon Press, New York (1999)

15. Macdonald, I.G.: Symmetric Functions and Hall Polynomials, 2nd edn. Oxford University Press, New York (1995)
16. Tavaré, S: Ancestral inference in population genetics In: Ecole d'Été de Probabilités de Saint Flour, Lecture Notes in Math. vol. 1837. Springer, Berlin (2004)
17. Griffiths, R.C., Tavaré, S.: Ancestral inference in population genetics. Statist. Sci. **9**, 307–319 (1994)
18. Forsythe, G.E., Leibler, R.A.: Matrix inversion by the Monte Carlo method. Math. Comp. **26**, 127–129 (1950)
19. Stephens, M., Donnelly, P.: Inference in molecular population genetics. J. R. Stat. Soc. Ser. B **62**, 605–635 (2000)
20. De Iorio, M., Griffiths, R.C.: Importance sampling on coalescent histories. I. Adv. Appl. Probab. **36**, 417–433 (2004)
21. Propp, J.G., Wilson, D.B.: Exact sampling with coupled Markov chains and applications to statistical mechanics. Random Struct. Algorithms **9**, 223–252 (1996)
22. Fearnhead, P.: Perfect simulation from population genetic models with selection. Theor. Popul. Biol. **59**, 263–279 (2001)
23. Mano, S.: Ancestral graph with bias in gene conversion. J. Appl. Probab. **50**, 239–255 (2013)
24. Liggett, T.M.: Stochastic Interacting Systems: Contact, Voter and Exclusion Processes. Springer, Berlin (1999)
25. Shimizu, A.: A measure valued diffusion process describing an n locus model incorporating gene conversion. Nagoya Math. J. **119**, 81–92 (1990)
26. Amari, S., Nagaoka, H.: Methods of Information Geometry. Translations of Mathematical Monograph vol. 191. Amer. Math. Soc. Providence (2000)
27. Levin, B., Reeds, J.: Compound multinomial likelihood functions are unimodal: proof of a conjecture of I.J. Good. Ann. Statist. **5**, 79–87 (1977)
28. Keener, R., Rothman, E., Starr, N.: Distribution of partitions. Ann. Statist. **15**, 1466–1481 (1978)
29. Hoshino, N.: Applying Pitman's sampling formula to microdata disclosure risk assessment. J. Official Statist. **17**, 499–520 (2001)
30. Takayama, N., Kuriki, S., Takemura, A.: A-hypergeometric distributions and Newton polytopes. Adv. in Appl. Math. **99**, 109–133 (2018)
31. Baayen, R.H.: Word Frequency Distribution. Kluwer, Dordrecht (2001)

Appendix A
Backgrounds on Related Topics

A.1   Symmetric Functions

This section presents minimum materials about symmetric functions used in this monograph. For a comprehensive discussion, see Macdonald's book [1] and Chap. 7 of Stanley's book [2].

Consider the ring $\mathbb{Z}[x_1, ..., x_k]$ of polynomials in independent variables $x_1, ..., x_k$ with rational integer coefficients. The symmetric group S_n acts on this ring by permutating the variables, and a polynomial is *symmetric* if it is invariant under this action. The symmetric polynomial form a subring

$$\Lambda_k = \mathbb{Z}[x_1, ..., x_k]^{S_k},$$

where Λ_k is a graded ring: we have $\Lambda_k = \oplus_{n \geq 0} \Lambda_k^n$, where Λ_k^n consists of the homogeneous symmetric polynomials of degree n, together with the zero polynomial. Let λ be a partition of length $l(\lambda) \leq k$. The polynomial

$$m_\lambda(x_1, ..., x_k) := \sum_\sigma \prod_{i=1}^{k} x_i^{\sigma_i}$$

summed over all distinct permutations σ of $\lambda = (\lambda_1, ..., \lambda_k)$ is called *monomial symmetric function*. The monomial symmetric functions such that $l(\lambda) \leq k$ and $|\lambda| = n$ form a basis of Λ_k^n. For example,

$$m_{(2,1)} = x_1^2 x_2 + x_1^2 x_3 + x_1^2 x_4 + x_1 x_2^2 + x_2^2 x_3 + x_2^2 x_4$$
$$+ x_1 x_3^2 + x_2 x_3^2 + x_3^2 x_4 + x_1 x_4^2 + x_2 x_4^2 + x_3 x_4^2 \in \Lambda_4^3.$$

For each $r \geq 1$ the r-th power sum is

© The Author(s) 2018
S. Mano, *Partitions, Hypergeometric Systems, and Dirichlet Processes in Statistics*,
JSS Research Series in Statistics, https://doi.org/10.1007/978-4-431-55888-0

$$p_r := m_{(r)} = \sum_{i=1}^{k} x_i^r.$$

The *power sum symmetric function* is defined as

$$p_\lambda := p_{\lambda_1} \cdots p_{\lambda_{l(\lambda)}} \in \Lambda_Q := \mathbb{Q}[p_1, p_2, \ldots].$$

Section I.4 of [1] discusses orthogonality among symmetric functions. The Schur symmetric function is defined as

$$s_\lambda(x) := \frac{\det(x_i^{\lambda_j + k - j})_{1 \le i, j \le k}}{\det(x_i^{k-j})_{1 \le i, j \le k}}.$$

It is well known that the Schur symmetric functions satisfy Cauchy's identity:

$$\prod_{1 \le i, j \le k} (1 - x_i y_j)^{-1} = \sum_{\{\lambda; l(\lambda) \le k\}} s_\lambda(x) s_\lambda(y). \tag{A.1}$$

In the theory of symmetric functions, the number of variables is usually irrelevant, provided that it is large enough, and it is often more convenient to work with symmetric functions in infinitely many variables. In the identity

$$\prod_{i,j} (1 - x_i y_j)^{-1} = \sum_\lambda s_\lambda(x) s_\lambda(y), \tag{A.2}$$

the sum is over all partitions. Let us introduce the orthonormality:

$$\langle s_\lambda, s_\mu \rangle = \delta_{\lambda, \mu}.$$

Here, $s_\lambda(x)$ such that $|\lambda| = n$ form an orthogonal basis of Λ^n, where Λ^n consists of homogeneous symmetric polynomials of degree n (refer to p. 18 of [1] for the definition). Using the power sum symmetric functions, the identity (A.2) is recast into

$$\prod_{i,j} (1 - x_i y_j)^{-1} = \sum_\lambda z_\lambda^{-1} p_\lambda(x) p_\lambda(y), \qquad z_\lambda := \prod_{i \ge 1} i^{c_i(\lambda)} c_i(\lambda)!,$$

and it follows that $\langle p_\lambda, p_\mu \rangle = \delta_{\lambda\mu} z_\lambda$, where p_λ form an orthogonal basis of Λ_Q.

The Jack symmetric function $P_\lambda^{(\alpha)}(x)$ is a generalization of the Schur symmetric function. Refer to Sect. VI.10 of [1] for the details. The Jack symmetric functions satisfy

$$\prod_{1 \le i, j \le k} (1 - x_i y_j)^{-1/\alpha} = \sum_\lambda (z_\lambda \alpha^{l(\lambda)})^{-1} p_\lambda(x) p_\lambda(y).$$

Next, we introduce the following orthogonality relation:

$$\langle p_\lambda, p_\mu \rangle_\alpha = \delta_{\lambda,\mu} z_\lambda \alpha^{l(\lambda)}. \tag{A.3}$$

The partial order among partitions of the same weight is defined as

$$\lambda \geq \mu \quad \Leftrightarrow \quad \lambda_1 + \cdots + \lambda_i \geq \mu_1 + \cdots + \mu_i, \qquad \forall i \geq 1,$$

for partitions μ and λ. It can be shown that (p. 322 of [1]) for each partition λ, there is a unique symmetric function $P_\lambda^{(\alpha)}$ such that

$$P_\lambda^{(\alpha)} = m_\lambda + \sum_{\mu < \lambda} u_{\lambda\mu}^{(\alpha)} m_\mu,$$

where

$$\langle P_\lambda^{(\alpha)}, P_\mu^{(\alpha)} \rangle_\alpha = 0, \qquad \lambda \neq \mu.$$

Here, the coefficient $u_{\lambda\mu}^{(1)}$ is called the Kostka number (see Sect. I.6 of [1]). The Jack symmetric functions $P_\lambda^{(\alpha)}$ such that $|\lambda| = n$ form an orthonormal basis of Λ^n. The inverse of the squared norm in the orthogonality relation (A.3) for each degree with normalization yields the Ewens sampling formula (2.12). In fact,

$$\sum_{\lambda \vdash n} \theta^{l(\lambda)} z_\lambda^{-1} = \sum_{\lambda \vdash n} \prod_{i=1}^n \left(\frac{\theta}{i}\right)^{c_i} \frac{1}{c_i!} = \frac{(\theta)_n}{n!}, \qquad \theta \equiv \frac{1}{\alpha},$$

and $n! \theta^{l(\lambda)} \{z_\lambda (\theta)_n\}^{-1}$ is the probability mass function of the Ewens sampling formula.

Remark A.1 The Jack symmetric function of $\alpha = 1$ is the Schur symmetric function, and that of $\alpha = 2$ with another normalization is known as the Zonal polynomial. The Zonal polynomial appears in integrations of the Haar measure of the orthogonal group, which appears in problems involving Wishart distributions [3]. Hashiguchi et al. discussed evaluation of the distribution function of the largest root of a Wishart matrix by using the holonomic gradient method discussed in Chap. 3 [4].

The Macdonald symmetric function is a further generalization of the Schur symmetric function. Chapter VI of [1] is devoted to this topic. The identity is

$$\prod_{i,j} \frac{(tx_i y_j; q)_\infty}{(x_i y_j; q)_\infty} = \sum_\lambda (z_\lambda(q,t))^{-1} p_\lambda(x) p_\lambda(y) \tag{A.4}$$

and the orthogonality relation is

$$\langle p_\lambda, p_\mu \rangle_{q,t} = \delta_{\lambda,\mu} z_\lambda(q,t), \tag{A.5}$$

where

$$z_\lambda(q, t) := z_\lambda \prod_{i \geq 1} \left(\frac{1 - q^i}{1 - t^i} \right)^{c_i}, \qquad (x; y)_n := \prod_{i=0}^{n-1} (1 - xy^i).$$

When $q = t$, the Macdonald symmetric function reduces to the Schur symmetric function, and when $q = 0$, it reduces to the Hall–Littlewood function. The Jack symmetric function appears in the limit $t = q^{1/\alpha}$, $q \to 1$.

The inverse of the squared norm in the orthogonality relation (A.5) for each degree with normalization yields a multiplicative measure induced by the exponential structure (2.16) with $w_i = (i - 1)!(t^i - 1)/(q^i - 1)$. Setting $x_1 = x$, $y_1 = 1$, and other variables to zero in the identity (A.4), we have

$$\frac{(tx; q)_\infty}{(x; q)_\infty} = \sum_{n=0}^{\infty} \sum_{\lambda \vdash n} (z_\lambda(q; t))^{-1} x^n.$$

From a q-analog of the negative binomial theorem (Theorem 12.2.5 in [5]):

$$_1\phi_0(t; -; q, x) := \sum_{n=0}^{\infty} \frac{(t; q)_n}{(q; q)_n} x^n = \frac{(tx; q)_\infty}{(x; q)_\infty},$$

we have

$$\sum_{\lambda \vdash n} (z_\lambda(q; t))^{-1} = \sum_{\lambda \vdash n} \prod_{i=1}^{n} \left(\frac{t^i - 1}{q^i - 1} \frac{1}{i} \right)^{c_i} \frac{1}{c_i!} = \frac{(t; q)_n}{(q; q)_n}. \tag{A.6}$$

A.2 Processes on Partitions

Stochastic processes on partitions and measure-valued processes are closely related. Shimizu [6] discussed a measure-valued diffusion taking values in probability measures on Young tableaux. The Dirichlet process is the reversible measure of a measure-valued diffusion called the Fleming–Viot process [7], which appeared as a model of genetic diversity. It is one of the most studied measure-valued processes, whose theory was founded by Feller [8]. This section presents minimum materials about the Fleming–Viot process used in this monograph. Chapter 10 of [9] is a detailed introduction. Further developments can be found in [10, 11], and in [12] in Japanese. Related issues such as coagulation and fragmentation are discussed in [13, 14]. The roles in modeling of genetic diversity can be found in [15].

Consider a diffusion process with a generator

$$L = \sum_{i,j} \frac{1}{2} a_{ij}(x) \frac{\partial^2}{\partial x_i \partial x_j} + \sum_i b(x) \frac{\partial}{\partial x_i} \tag{A.7}$$

whose backward equation for the transition density ϕ has the form $\partial\phi/\partial t = L\phi$. The forward equation is $\partial\phi/\partial t = L^+\phi$, where L^+ is the adjoint operator of L:

$$L^+ \bullet \phi = \sum_{i,j} \frac{1}{2} \frac{\partial^2}{\partial x_i \partial x_j} (a_{ij}(x)\phi) - \sum_i \frac{\partial}{\partial x_i} (b(x)\phi).$$

For a test function f and a probability measure μ, let us introduce a notation $\langle f, \mu \rangle := \int f(x)d\mu(x)$. Let us assume existence of the unique stationary measure π for the diffusion. It should satisfy $L^+\pi = 0$, since

$$0 = \frac{d}{dt}\langle f, \pi \rangle = \langle Lf, \pi \rangle = \langle f, L^+\pi \rangle, \qquad \forall f. \tag{A.8}$$

Moreover, if π is *reversible*, π should satisfy

$$\langle Lf, g\pi \rangle = \langle Lg, f\pi \rangle, \qquad \forall f, g.$$

It can be observed that this condition is equivalent to $L_j^+\pi = 0$ for $\forall j$, where

$$L_j^+ \bullet \phi = \sum_i \frac{1}{2} \frac{\partial}{\partial x_j} (a_{ij}(x)\phi) - b_j(x)\phi, \qquad L^+ = \sum_j \frac{\partial}{\partial x_j} \bullet L_j^+.$$

Let us consider a diffusion process whose diffusion and drift coefficients are given by $a_{ij}(x) = x_i(\delta_{ij} - x_j)$ and $b_i(x) = \alpha(\sum_j x_j - mx_i)/2, \alpha > 0$, respectively. Consider a generator

$$L = \sum_{i=1}^m \sum_{j=1}^m \frac{x_i(\delta_{ij} - x_j)}{2} \frac{\partial^2}{\partial x_i \partial x_j} + \frac{\alpha}{2} \sum_{i=1}^m \left(\sum_{j=1}^m x_j - mx_i \right) \frac{\partial}{\partial x_i} \tag{A.9}$$

in the state space $(x_1, ..., x_m) \in \Delta_{m-1}$. A diffusion process in the simplex with covariance diffusion coefficients is called *Wright–Fisher diffusion*. Since L^+ and L_j^+ annihilate the density (4.3) of the symmetric m-variate Dirichlet distribution of parameter α, the Dirichlet distribution is the reversible measure of the Wright–Fisher diffusion.

Remark A.2 The condition that L_j^+ annihilates the density is useful to obtain explicit expressions of the density of the reversible measure. A demonstration is given in [16]. Moreover, the condition is useful to construct a sampler from random partitions (see Sect. 5.1.4).

Remark A.3 For the Wright–Fisher diffusion (A.9), Griffiths [17] obtained an expansion of the transition density in terms of orthogonal polynomials of the form

$$f(x, y; t) = \pi_\alpha(y) \left\{ 1 + \sum_{i \geq 1} P_i(x) P_i(y) \exp\left(-\frac{i(i-1+m\alpha)}{2} t \right) \right\},$$

where π_α is the density of the symmetric m-variate Dirichlet distribution and $\{P_i(x), i \in \mathbb{N}\}$ are orthonormal Jacobi polynomials on the m-variate symmetric distribution scaled such that $\mathbb{E}_{\pi_\alpha}[P_i(X)P_j(X)] = \delta_{i,j}, i, j \in \mathbb{N}$. The symmetric kernel reflects the reversibility of the process.

Taking the monomial

$$q_n(x) = \frac{n!}{n_1! \cdots n_m!} x^n, \qquad x^n := \prod_{i=1}^m x_i^{n_i}, \tag{A.10}$$

as a test function, we obtain the Dirichlet-multinomial distribution (4.10):

$$p(n) := \langle q_n, \pi_\alpha \rangle = \binom{-m\alpha}{n}^{-1} \prod_{i=1}^m \binom{-\alpha}{n_i}.$$

The Dirichlet-multinomial distribution is an EPPF introduced in Sect. 4.3. The stationarity condition (A.8) yields the recurrence relation

$$p(n) = \frac{n-1}{m\alpha + n - 1} \sum_{i=1}^m \frac{n_i - 1}{n-1} p(n - e_i)$$

$$+ \frac{\alpha}{m\alpha + n - 1} \sum_{i=1}^m \sum_{j=1}^m \frac{n_j + 1 - \delta_{ij}}{n} p(n - e_i + e_j) \tag{A.11}$$

with the boundary condition $p(e_i) = 1/m, i \in [m]$. Taking the limit $m \to \infty, \alpha \to 0$ with $\theta \equiv m\alpha$ in the Dirichlet-multinomial distribution gives the Ewens sampling formula (2.12) (Remark 4.6). Rewriting (A.11) in terms of size indices (see exponential structures in Sect. 2.1) and taking the limit, we have

$$\mu_n(c) = \frac{n-1}{\theta + n - 1} \sum_{i=1}^{n-1} \frac{i(c_i + 1)}{n-1} \mu_{n-1}(c + e_i - e_{i+1})$$

$$+ \frac{\theta}{\theta + n - 1} \left\{ \frac{c_1}{n} \mu_n(c) + \sum_{i=2}^n \frac{i(c_i + 1)}{n} \mu_n(c - e_{i-1} + e_i - e_1) \right\} \tag{A.12}$$

with the boundary condition $\mu_1(e_1) = 1$. The Ewens sampling formula satisfies this recurrence relation.

The above observation implies that the Dirichlet process is the reversible measure of an infinite-dimensional diffusion. Such a diffusion was formulated by Fleming and Viot [7], which is now called a Fleming–Viot diffusion. Let E be a compact metric space. Let $\mathscr{C}(E)$ be set of continuous real-valued functions on E and $\mathscr{P}(E)$ be the family of Borel probability measures on E. For $f \in \mathscr{B}(E)$ and $\mu \in \mathscr{P}(E)$, define

$$\phi(\mu) = F(\langle f_1, \mu \rangle, ..., \langle f_k, \mu \rangle) \in \mathscr{C}(\mathscr{P}(E))$$

for some $k \in \mathbb{N}$. A generator of the Fleming–Viot diffusion with a linear operator B on $\mathscr{C}(E)$ is defined as

$$G\phi(\mu) = \frac{1}{2} \sum_{i,j=1}^{k} (\langle f_i f_j, \mu \rangle - \langle f_i, \mu \rangle \langle f_j, \mu \rangle) F_{,ij}(\langle f_1, \mu \rangle, \cdots, \langle f_k, \mu \rangle)$$

$$+ \sum_{i=1}^{k} \langle Bf_i, \mu \rangle F_{,i}(\langle f_1, \mu \rangle, \cdots, \langle f_k, \mu \rangle), \qquad (A.13)$$

where $F_{,i}(x_1, ..., x_k) = \partial F / \partial x_i$.

Example A.1 (Dirichlet distribution) Let $E = \{1/m, 2/m, ..., 1\}$ and define

$$Bf_i = \frac{\alpha}{2} \sum_{j=1}^{m} (f_j - f_i).$$

The solution of the martingale problem defined by the generator (A.13) is $\mu(t) = \sum_{i=1}^{m} x_i(t) \delta_{i/m}$, where $x(t)$ follows the Wright–Fisher diffusion governed by the generator (A.9). The transition density is given in Remark A.3 and the reversible measure is π_α.

Example A.2 (Dirichlet process) Let $E = [0, 1]$ and define

$$Bf(x) = \frac{\theta}{2} \int_0^1 \{f(y) - f(x)\} dy.$$

The reversible measure of the diffusion governed by the generator (A.13) has the form $\mu = \sum_{i \geq 1} x_i \delta_{V_i}$, $V_i \sim \text{Unif.}([0, 1])$, where x follows the GEM distribution (Remark 4.3). Therefore, μ follows the Dirichlet process $\text{DP}(\theta; \text{Unif.}(E))$, which appeared in Sect. 4.3.

The Ewens sampling formula is a random integer partition and a sample from the Dirichlet process. We have interest in stochastic processes on partitions which is related with the Fleming–Viot diffusion. Kingman discovered such a $\mathscr{P}_{\mathbb{N}}$-valued process, which is called *Kingman's coalescent*. It is a Markov chain on partitions with the following transition rule. Assume that the process is in the state $\{A_1, ..., A_l\}$. The only possible transitions are one of the $l(l-1)/2$ partitions obtained by merging parts A_i and A_j to form $A_i \cup A_j$ and leaving all other parts uncharged at rate one. The length of partition $(L_t; t \geq 0)$, $L_0 = n$, follows the death process whose transitions are $l \to l - 1$ at rate $l(l-1)/2$. The process is eventually absorbed into the state of one. Let us consider this process as generating a tree upward from the leaves to the root. Time is vertical, and parts at a given time are located along a horizontal line. A merger is called *coalescence*. Let us introduce a Poisson process of marks, which is called *mutation*, along all branches of this tree at rate $\theta/2$ per unit length. Then, a

Fig. A.1 A realization of the coalescent tree of the partition (2, 2, 1)

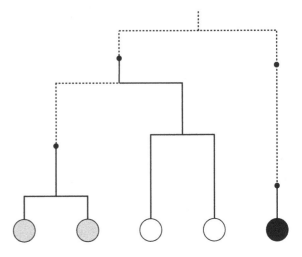

random integer partition is generated by the equivalence relation $i \sim j$ if there is no mutation on the unique path in the tree that joins i to j, see Fig. A.1.

The relationship between the Fleming–Viot diffusion and the Kingman's coalescent can be understood by the notion of *duality* between Markov chains. The method of duality has been widely used in analyses of infinite particle systems. Many examples of the use can be found in [19]. If $(X_t; t \geq 0)$ with $X_0 = x$ and $(Y_t; t \geq 0)$ with $Y_0 = y$ are Markov processes in state spaces E_x and E_y, respectively, then the processes X_t and Y_t are said to be dual with respect to a kernel $k(x, y)$ if the identity

$$\mathbb{E}_x(k(X_t, y)) = \mathbb{E}_y(k(x, Y_t)), \qquad \forall x \in E_x, \ y \in E_y \qquad (A.14)$$

holds. Consider generators of G_x for x_t and G_y for y_t. Then, the duality relationship (A.14) will be satisfied if the identity $G_x k(x, y) = G_y k(x, y)$ holds for all $x \in E_x$ and $y \in E_y$. Therefore, if we know G_x, we may identify the dual G_y. Let us consider the Wright–Fisher diffusion governed by the generator (A.9) and take the kernel $k(x, n) = x^n / \mathbb{E}_{\pi_\alpha}(x^n)$. We observe

$$Lk(x, n) = -\frac{n(n-1+m\alpha)}{2}k(x, n) + \frac{n(n-1+m\alpha)}{2}\sum_{i=1}^{m}\frac{n_i}{n}k(x, n - e_i) = G_n k(x, n),$$

and can lead the process $(N_t; t \geq 0)$, $N_0 = n$ from this expression. Taking the limit $m \to \infty$, $\alpha \to 0$ with $\theta \equiv m\alpha$, we can observe that N_t follows the infinite-dimensional death process whose transitions are $l \to l - e_i$ at late $l(l - 1 + \theta)/2 \times l_i/l$. This process is certainly generated by Kingman's coalescent; the rate $l(l - 1)/2$ comes from the coalescence, while the rate $l\theta/2$ comes from the mutation.

Remark A.4 An extension of Kingman's coalescent, the Λ-coalescent, has been extensively studied. The fairly recent surveys are [20, 21]. The transition rule is as follows. Assume that the process is in the state $\{A_1, ..., A_b\}$. Then, k blocks merge with rates

$$\lambda_{b,k} = \int_0^1 x^{k-2}(1-x)^{b-k} \Lambda(dx), \quad 2 \le k \le b,$$

where Λ is a nonnegative finite measure on $[0, 1]$. Pitman [22] showed that the array of rates $(\lambda_{b,k})$ is consistent if and only if $\lambda_{b,k} = \lambda_{b+1,k} + \lambda_{b+1,k+1}$. This condition originates from the de Finetti's representation theorem (Theorem 1.1), because the representation can be regarded as infinite exchangeable sequences of binomial random variables. Kingman's coalescent is the case of $\Lambda = \delta_0$. Another well-investigated example is the beta coalescent, whose Λ is the density of Beta$(2 - \alpha, \alpha)$, $\alpha \in (0, 2)$. The diffusion and jump part of the generator of a Λ-Fleming–Viot process are

$$G\phi(\mu) = \Lambda(\{0\})G_0\phi(\mu)$$
$$+ \int_{(0,1]} \int_E \{\phi((1-x)\mu + x\delta_a) - \phi(\mu)\} \mu(da) \frac{\Lambda_0(dx)}{x^2},$$

where G_0 is the diffusion term in (A.13) and Λ_0 is the Λ on $(0, 1]$.

References

1. Macdonald, I.G.: Symmetric Functions and Hall Polynomials, 2nd edn. Oxford University Press, New York (1995)
2. Stanley, R.P.: Enumerative Combinatorics, vol. 2. Cambridge University Press, New York (1999)
3. Takemura, A.: Zonal Polynomials. Lecture Notes-Monograph Series, vol. 4. Michigan State University (1984)
4. Hashiguchi, H., Numata, Y., Takayama, N., Takemura, A.: The holonomic gradient method for the distribution function of the largest root of a Wishart matrix. J. Multivar. Anal. **117**, 296–312 (2013)
5. Ismail, M.E.H.: Classical and Quantum orthogonal polynomials in one variable. In: Encyclopedia of Mathematics and its Applications, vol. 98. Cambridge University Press, New York (2005)
6. Shimizu, A.: Stationary distribution of a diffusion process taking values in probability distributions on the partitions. In: Kimura, M., Kallianpur, G., Hida, T. (eds.) Stochastic Methods in Biology. Lecture Notes in Biomathematics, vol. 70, pp. 100–114. Springer, Berlin (1987)
7. Fleming, W.H., Viot, M.: Some measure valued Markov processes in population genetics theory. Indiana Univ. Math. J. **28**, 817–843 (1979)
8. Feller, W.: Diffusion processes in genetics. In: Proceedings of the Second Berkeley Symposium on Mathematical Statistics and Probability, pp. 227–246. University of Carfornia Press, Berkeley (1950)
9. Ethier, S.N., Kurtz, T.G.: Markov Processes: Characterization and Convergence. Wiley, New Jersey (1986)
10. Dawson, D.A.: Measure-valued Markov processes. In: Ecole d'Été de Probabilités de Saint Flour. Lecture Notes in Mathematics, vol. 1541. Springer, Berlin (1993)
11. Etheridge, A.M.: An Introduction to Superprocess. University Lecture Notes, vol. 20. American Mathematical Society, Providence (2000)

12. Handa, K.: Generalizations of Wright-Fisher diffusions and related topics. Proc. Inst. Stat. Math. **60**, 327–339 (2012)
13. Pitman, J.: Combinatorial Stochastic Processes. In: Ecole d'Été de Probabilités de Saint Flour, Lecture Notes in Mathematics, vol. 1875. Springer, Berlin (2006)
14. Bertoin, J.: Random Fragmentation and Coagulation Process. Cambridge University Press, Cambridge (2006)
15. Durrett, R.: Probability Models for DNA Sequence Evolution. Springer, New York (2008)
16. Handa, K.: Reversible distributions of multi-allelic Gillespie-Sato diffusion models. Ann. Inst. H. Poincaré Prob. Stat. **40**, 569–597 (2004)
17. Griffiths, R.C.: A transition density expansion for a multi-allele diffusion model. Adv. Appl. Probab. **11**, 310–325 (1979)
18. Kingman, J.F.C.: The coalescent. Stoch. Process Appl. **13**, 235–248 (1982)
19. Liggett, T.M.: Stochstic Interacting Systems: Contact, Voter and Exclusion Processes. Springer, Berlin (1999)
20. Berestycki, N.: Recent progress in coalescent theory. Ensaios Mathemáticos **16**, 1–193 (2009)
21. Birkner, M. Blath, J.: Measure-valued diffusions, general coalescents and population genetic inference. In: Blath, J., Mörters, P., Scheutzow, M. (eds.) Trends in Stochastic Analysis. London Mathematical Society Lecture Note Series, vol. 353, pp. 329–363. Cambridge University Press, Cambridge (2009)
22. Pitman, J.: Coalescent with multiple collisions. Ann. Probab. **27**, 1870–1902 (1999)

Index

© The Author(s) 2018
S. Mano, *Partitions, Hypergeometric Systems, and Dirichlet Processes in Statistics*,
JSS Research Series in Statistics, https://doi.org/10.1007/978-4-431-55888-0

Printed in the United States
By Bookmasters